Lecture Notes in Computer Science 6790

Commenced Publication in 1973
Founding and Former Series Editors:
Gerhard Goos, Juris Hartmanis, and Jan van Leeuwen

W0192936

Abdelkader Hameurlain   Josef Küng
Roland Wagner (Eds.)

Transactions on Large-Scale Data- and Knowledge-Centered Systems III

Special Issue on Data and Knowledge Management
in Grid and P2P Systems

 Springer

Editors-in-Chief

Abdelkader Hameurlain
Paul Sabatier University
Institut de Recherche en Informatique de Toulouse (IRIT)
118, route de Narbonne, 31062 Toulouse Cedex, France
E-mail: hameur@irit.fr

Josef Küng
Roland Wagner
University of Linz, FAW
Altenbergerstraße 69
4040 Linz, Austria
E-mail: {j.kueng,rrwagner}@faw.at

ISSN 0302-9743 (LNCS)   e-ISSN 1611-3349 (LNCS)
ISSN 1869-1994 (TLDKS)
ISBN 978-3-642-23073-8   ISBN 978-3-642-23074-5 (eBook)
DOI 10.1007/978-3-642-23074-5
Springer Heidelberg Dordrecht London New York

Library of Congress Control Number: 2011933662

CR Subject Classification (1998): H.2, I.2.4, H.3, H.4, J.1, H.2.8

Typesetting: Camera-ready by author, data conversion by Scientific Publishing Services, Chennai, India

Printed on acid-free paper

Springer is part of Springer Science+Business Media (www.springer.com)

Preface

The LNCS journal Transactions on Large-Scale Data- and Knowledge-Centered Systems focuses on data management, knowledge discovery, and knowledge processing, which are core and hot topics in computer science. Since the 1990s, the Internet has become the main driving force behind application development in all domains. An increase in the demand for resource sharing across different sites connected through networks has led to an evolution of data- and knowledge-management systems from centralized systems to decentralized systems enabling large-scale distributed applications providing high scalability. Current decentralized systems still focus on data and knowledge as their main resource. Feasibility of these systems relies basically on P2P (peer-to-peer) techniques and the support of agent systems with scaling and decentralized control. Synergy between Grids, P2P systems, and agent technologies is the key to data- and knowledge-centered systems in large-scale environments.

The third issue of this journal focuses on data and knowledge management in grid and P2P systems. The importance of these systems arises from their main characteristics: the autonomy and the dynamics of peers, the decentralization of control, and the effective and transparent sharing of large-scale distributed resources.

This special issue of TLDKS contains two kinds of papers. First, it contains a selection of the best papers from the third International Conference on Data Management in Grid and Peer-to-Peer Systems (Globe 2010), which was held September 1–2, 2010 in Bilbao, Spain. The authors were invited to submit extended versions for a new round of reviewing. Second, it contains, after reviewing process, a selection of 6 papers from 18 submitted papers in response to the call for papers for this issue.

The content of this special issue, which is centered on data and knowledge management in grid and P2P systems, covers a wide range of different and hot topics in the field, mainly: replication, semantic web, information retrieval, data storage, source selection, and large-scale distributed applications.

We would like to express our thanks to the external reviewers and editorial board for thoroughly refereeing the submitted papers and ensuring the high quality of this special issue. Special thanks go to Gabriela Wagner for her valuable work in the realization of this edition.

May 2011

Abdelkader Hameurlain
Josef Küng
Roland Wagner

Editorial Board

Table of Contents

Replication in DHTs Using Dynamic Groups*

Reza Akbarinia[1], Mounir Tlili[2], Esther Pacitti[1],
Patrick Valduriez[1], and Alexandre A.B. Lima[3]

[1] INRIA and LIRMM, Montpellier, France
[2] INRIA and LINA, Univ. Nantes, France
[3] COPPE/UFRJ, Rio de Janeiro, Brazil
{Reza.Akbarinia,Patrick.Valduriez}@inria.fr,
Mounir.Tlili@univ-nantes.fr, pacitti@lirmm.fr, assis@cos.ufrj.br

Abstract. Distributed Hash Tables (DHTs) provide an efficient solution for data location and lookup in large-scale P2P systems. However, it is up to the applications to deal with the availability of the data they store in the DHT, e.g. via replication. To improve data availability, most DHT applications rely on data replication. However, efficient replication management is quite challenging, in particular because of concurrent and missed updates. In this paper, we propose a complete solution to data replication in DHTs. We propose a new service, called Continuous Timestamp based Replication Management (CTRM), which deals with the efficient storage, retrieval and updating of replicas in DHTs. In CTRM, the replicas are maintained by groups of peers which are determined dynamically using a hash function. To perform updates on replicas, we propose a new protocol that stamps the updates with timestamps that are generated in a distributed fashion using the dynamic groups. Timestamps are not only monotonically increasing but also continuous, i.e. without gap. The property of monotonically increasing allows applications to determine a total order on updates. The other property, i.e. continuity, enables applications to deal with missed updates. We evaluated the performance of our solution through simulation and experimentation. The results show its effectiveness for replication management in DHTs.

1 Introduction

Distributed Hash Tables (DHTs) such as CAN [17], Chord [21] and Pastry [20], provide an efficient solution for data location and lookup in large-scale P2P systems. While there are significant implementation differences between DHTs, they all map a given key k onto a peer p using a hash function and can lookup p efficiently, usually in $O(log\ n)$ routing hops, where n is the number of peers [5]. One of the main characteristics of DHTs (and other P2P systems) is the dynamic behavior of peers which can join and leave the system frequently, at any time. When a peer gets offline, its data becomes unavailable. To improve data availability, most applications which are built on top of DHTs rely on data replication by storing the (*key, data*) pairs at several peers, *e.g.* using several hash functions. If one peer is unavailable, its data can still be

---

* Work partially funded by the ANR DataRing projet.

A. Hameurlain, J. Küng, and R. Wagner (Eds.): TLDKS III , LNCS 6790, pp. 1–19, 2011.

retrieved from the other peers that hold a replica. However, update management is difficult because of the dynamic behaviour of peers and concurrent updates. There may be *replica holders* (i.e. peers that maintain replicas) that do not receive the updates, e.g. because they are absent during the update operation. Thus, we need a mechanism that efficiently determines whether a replica on a peer is up-to-date, despite missed updates. In addition, to deal with concurrent updates, we need to determine a total order on the update operations.

In this paper, we give an efficient solution to replication management in DHTs. We propose a new service, called Continuous Timestamp based Replication Management (CTRM), which deals with the efficient storage, retrieval and updating of replicas in DHTs. In CTRM, the replicas are maintained by groups of peers, called replica holder groups, which are dynamically determined using a hash function. To perform updates on replicas, we propose a new protocol that stamps the updates with timestamps that are generated in a distributed fashion using the members of the groups. The updates' timestamps are not only monotonically increasing but also continuous, i.e. without gap. The property of monotonically increasing allows CTRM to determine a total order on updates and to deal with concurrent updates. The continuity of timestamps enables replica holders to detect the existence of missed updates by looking at the timestamps of the updates they have received. Examples of applications that can take advantage of continuous timestamping are the P2P collaborative text editing applications, e.g. P2P Wiki [23], which need to reconcile the updates done by collaborating users. We analyze the network cost of CTRM using a probabilistic approach, and show that its cost is very low in comparison to two baseline services in DHTs. We evaluated CTRM through experimentation and simulation; the results show its effectiveness. In our experiments, we compared CTRM with two baseline services, and the results show that with a low overhead in update response time, CTRM supports fault-tolerant data replication using continuous timestamps. The results also show that data retrieval with CTRM is much more efficient than the baseline services. We investigated the effect of peer failures on the correctness of CTRM and the results show that it works correctly even in the presence of peer failures.

This paper is an extended version of [2] that involves at least 38% of new material including the following contributions. First, in Section 4, we extend the concept of replica holder groups which are essential for our solution. In particular, we deal with the dynamic behaviour of the group members, which can leave the system at any time. Second, in Section 6, we give a communication cost analysis of our solution, using a probabilistic approach, and compare the cost of our solution with those of two baseline services. We also include more discussion in Section 8 about related work on replication management in P2P systems.

The rest of this paper is organized as follows. In Section 2, we define the problem we address in this paper. In Section 3, we give an overview of our CTRM service and its operations. In Section 4, we describe the replica holder groups in CTRM. In Section 5, we propose the new UCT protocol which is designed for updating replicas in CTRM. Section 6 presents a cost analysis of the CTRM service. Section 7 reports a performance evaluation of our solution. Section 8 discusses related work, and Section 9 concludes.

2 Problem Definition

In this paper we deal with improving data availability in DHTs. Like several other protocols and applications designed over DHTs, e.g. [5], we assume that the lookup service of the DHT behaves properly. That is, given a key k, it either finds correctly the responsible for k or reports an error, *e.g.* in the case of network partitioning where the responsible peer is not reachable.

To improve data availability, we replicate each data at a group of peers of the DHT which we call *replica holders*. Each replica holder keeps a *replica copy* of a replicated data. Each replica may be updated locally by a replica holder or remotely by other peers of the DHT. This model is in conformance with the *multi-master replication* model [15].

The problem that arises is that a replica holder may fail or leave the system at any time. Thus, the replica holder may miss some updates during its absence. Furthermore, updates on different replicas of a data may be performed in parallel, i.e. concurrently. To ensure consistency, updates must be applied to all replicas in a specific total order.

In this model, to ensure eventual consistency of replicas, we need a distributed mechanism that determines 1) a total order for the updates; 2) the number of missed updates at a replica holder. Such a mechanism allows dealing with concurrent updates, i.e. committing them in the same order at all replica holders. In addition, it allows a rejoining (recovering) replica holder to determine whether its local replica is up-to-date or not, and how many updates should be applied on the replica if it is not up-to-date.

In this paper, we aim at developing a replication management service supporting the above-mentioned mechanism in DHTs. One solution for realizing such a mechanism is to stamp the updates with timestamps that are monotonically increasing and continuous. We call such a mechanism *update with continuous timestamps*.

Let *patch* be the action (or set of actions) generated by a peer during one update operation. Then, the property of update with continuous timestamps can be defined as follows.

Definition 1: Update with continuous timestamps (UCT). An update mechanism is UCT *iff* : the update patches are stamped by increasing real numbers such that the difference between the timestamps of any two consecutive committed updates is one.

Formally, consider two consecutive committed updates u_1 and u_2 on a data d, and let pch_1 and pch_2 be the patches of u_1 and u_2, respectively. Assume that u_2 is done after u_1, and let t_1 and t_2 be the timestamps of pch_1 and pch_2 respectively. Then we should have $t_2 = t_1 + 1$;

To support the UCT property in a DHT, we must deal with two challenges: 1) To generate continuous timestamps in the DHT in a distributed fashion; 2) To ensure that any two consecutive generated timestamps are used for two consecutive updates. Dealing with the first challenge is hard, in particular due to the dynamic behavior of peers, which can leave or join the system at any time and frequently. This behavior makes inappropriate the timestamping solutions based on physical clocks, because the distributed clock synchronization algorithms do not guarantee good synchronization precision if the nodes are not linked together long enough [16]. Addressing the second challenge is difficult as well, because there may be generated timestamps which are used for no update, e.g. because the timestamp requester peer may fail before doing the update.

3 Overview of Replication Management in CTRM

CTRM (Continuous Timestamp based Replication Management) is a replication management service which we designed to deal with efficient storage, retrieval and updating of replicas on top of DHTs, while supporting the UCT property.

To provide high data availability, CTRM replicates each data in the DHT at a group of peers, called *replica holder group*. For each replicated data, there is a replica holder group which is determined dynamically by using a hash function. To know the group which holds the replica of a data, peers of the DHT apply the hash function on the data ID, and using the DHTs lookup service to find the group. The details of the replica holder groups are presented in Section 4.

3.1 Data Update

CTRM supports multi-master data replication, i.e. any peer in the DHT can update the replicated data. After each update on a data by a peer p, the corresponding patch, i.e. set of update actions, is sent by p to the replica holder group where a monotonically increasing timestamp is generated by one of the members, i.e. the responsible for the group. Then the patch and its timestamp are published to the members of the group using an update protocol, called UCT protocol. The details of the UCT protocol are presented in Section 5.

3.2 Replica Retrieval

To retrieve an up-to-date replica of a data, the request is sent to the peer that is responsible for the data's replica holder group. The responsible peer sends the data and the latest generated timestamp to the group members, one by one. The first member that maintains an up-to-date replica returns it to the requester. To check whether their replicas are up-to-date, replica holders check the two following conditions, called *up-to-date conditions*:

1. The latest generated timestamp is equal to the timestamp of the latest patch received by the replica holder.
2. The timestamps of the received patches are continuous, i.e. there is no missed update.

The UCT protocol, which is used for updating the data in CTRM, guarantees that if at peer p there is no gap between the timestamps and the last timestamp is equal to the last generated one, then p has received all replica updates. In contrast, if there is some gap in the received timestamps, then there should be some missed updates at p.

If during the replica retrieval operation, a replica holder p understands that it misses some updates, then it retrieves the missed updates and their timestamps from the group's responsible peer or other members that hold them, and updates its replica.

In addition to the replica retrieval operation, the up-to-date conditions are also verified periodically by each member of the group. If the conditions do not hold, the member updates its replica by retrieving the missed updates from other members of the group.

4 Replica Holder Groups

Replica holder groups are dynamic groups of peers which are responsible for maintaining the replicas of data, timestamping the updates, and returning up-to-date data to the users.

In this section, we first describe the idea behind the replica holder groups, then discuss on how they assure their correct functionality in the presence of peer join/departures, which can be frequent in P2P systems.

4.1 Basic Ideas

Let G_k be the group of peers that maintain the replicas of a data whose ID is k. We call these peers the *replica holder group of* k. For each group, there is a responsible peer which is also one of its members. For choosing a responsible peer for the group G_k, we use a hash function h_r, and the peer p that is responsible for key=$h_r(k)$ in the DHT, is the responsible for G_k. In this paper, the peer that is responsible for key=$h_r(k)$ is denoted by $rsp(k , h_r)$, *i.e.* called responsible for k with regard to hash function h_r. In addition to $rsp(k , h_r)$, some of the peers that are close to it, .e.g. its neighbors, are members of G_k. Each member of the group knows the address of other members of the group. The number of members of a replica holders group, i.e. $|G_k|$, is a system's parameter.

Each group member p periodically sends alive messages to the group's responsible peer, and the responsible peer returns to it the current list of members. If the responsible peer does not receive an alive message from a member, it assumes that the member has failed. When a member of a group leaves the system or fails, after getting aware of this departure, the responsible peer invites a close peer to join the group, *e.g.* one of its neighbors. The new member receives from the responsible peer a list of other members as well as up-to-date replicas of all data replicated by the group.

The peer p that is responsible for G_k generates timestamps for the updates done on the data k. For generating the timestamps, it uses a local counter called *counter of* k *at* p which we denote as $c_{p,k}$. When p receives an update request for a data k, it increments the value of $c_{p,k}$ by one and stores the update patch and the timestamp over the other members of the group using a protocol which we describe in Section 5.

In the situations where the group's responsible peer leaves the system or fails, another peer takes it over. This responsibility change can also happen in the situations where another peer joins the system and becomes responsible for the key $h_r(k)$ in the DHT. In the next section, we discuss on how the responsibility migrates in these situations, and how the new responsible peer initializes its counter to the correct timestamp value, i.e. to the value of the last generated timestamp.

4.2 Dealing with Departure of Group's Responsible Peer

The responsible peer is the most important member of the group. In the management of the replica groups, we must deal with the cases where the responsible peer leaves the system or fails. The main issues are: how to determine the next responsible peer, and how to initialize the counter values on it.

4.2.1 Who Is the Next Group's Responsible?

As discussed in Section 4.1, the responsible for the group G_k is the peer that is responsible for the key $h_r(k)$ in the DHT. Notice that at any time, there is a responsible peer for each key. If the current responsible for the key $h_r(k)$ leaves the DHT, another peer, say p, becomes responsible for the key. This peer p becomes also the new responsible for the group G_k. Therefore, if a peer wants to contact the responsible for G_k, the lookup service of the DHT gives it the address of p.

An interesting question is about the relationship between the current and the next responsible peer in the DHT. To answer the question, we observe that, in DHTs, the next peer that obtains the responsibility for k is typically a neighbor of the current responsible peer, so the next responsible peer is one of the members of the group. We now illustrate this observation with CAN and Chord, two popular DHTs.

Let $rsp(k, h_r)$ be the current responsible peer for group G_k , and $nrsp(k, h_r)$ be the one that takes it over. Let us assume that peer q is $rsp(k, h_r)$ and peer p is $nrsp(k, h_r)$. In CAN and Chord, there are only two ways by which p would obtain the responsibility for k. First, q leaves the P2P system or fails, so the responsibility of k is assigned to p. Second, p joins the P2P system which assigns it the responsibility for k, so q looses the responsibility for k despite its presence in the P2P system. In both cases, we show that both CAN and Chord have the property that $nrsp(k, h_r)$ is one of the neighbors $rsp(k, h_r)$.

Chord. In Chord [21], each peer has an m-bit identifier (ID). The peer IDs are ordered in a circle and the neighbors of a peer are the peers whose distance from p clockwise in the circle is 2^i for $0 \le i \le m$. The responsible for $h_r(k)$ is the first peer whose ID is equal or follows $h_r(k)$. Consider a new joining peer p with identifier ID_p. Suppose that the position of p in the circle is just between two peers p_1 and p_2 with identifiers ID_1 and ID_2, respectively. Without loss of generality, we assume that $ID_1 < ID_2$, thus we have $ID_1 < ID_p < ID_2$. Before the entrance of p, the peer p_2 was

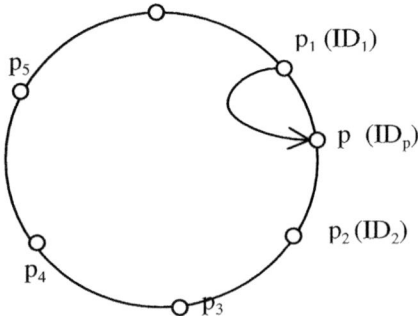
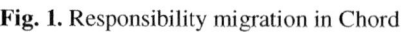
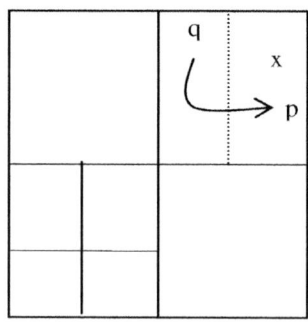

Fig. 1. Responsibility migration in Chord

Fig. 2. Responsibility migration in CAN, based on a two dimensional coordinate space

responsible for k if and only if $ID_1 < h_r(k) \le ID_2$. When p joins Chord, it becomes responsible for k if and only if $ID_1 < h_r(k) \le ID_p$ (see Figure 1). In other words, p becomes responsible for a part of the keys for which p_2 was responsible. Since the distance clockwise from p to p_2 is 2^0, p_2 is a neighbor of p. Thus, in the case of join, the next responsible peer is one of the neighbors of the current responsible. If p leaves the system or fails, the next peer in the circle, say p_2, becomes responsible for its keys.

CAN. We show this property by giving a brief description of CAN's protocol for joining and leaving the system [17]. CAN maintains a virtual coordinate space partitioned among the peers. The partition which a peer owns is called its zone. According to CAN, a peer p is responsible for $h_r(k)$ if and only if $h_r(k)$ is in p's zone. When a new peer, say p, wants to join CAN, it chooses a point X and sends a join request to the peer whose zone involves X. The current owner of the zone, say q, splits its zone in half and the new peer occupies one half, then q becomes one of p's neighbors (see Figure 2). Thus, in the case of join, $nrsp(k, h_r)$ is one of the neighbors of $rsp(k, h_r)$. Also, when a peer p leaves the system or fails, its zone will be occupied by one of its neighbors, *i.e.* the one that has the smallest zone. Thus, in the case of leave or fail, $nrsp(k, h_r)$ is one of the neighbors of $rsp(k, h_r)$, and that neighbor is known for $rsp(k, h_r)$.

Following the above discussion, in Chord or CAN when the current group's responsible peer leaves the system or fails, one of its neighbors becomes the next responsible peer.

4.2.2 Timestamp Initialization

In the case where a responsible peer q leaves the system or fails, the next responsible p should initialize its counters to the last value of generated timestamps. In CTRM, we consider two different situations for counter initialization: 1) normal departure of q; 2) failure of q.

Normal Departure

When a responsible peer leaves the system normally, i.e. without failure, the counter initialization is done by directly transferring the counters from the current responsible peer to the next one at the end of its responsibility.

Let q and p be two peers, and $K' \subseteq K$ be the set of keys for which q is the current responsible peer, and p is the next responsible. Once q reaches the end of its responsibility for the keys in K', *e.g.* before leaving the system, it sends to p all its counters that have been initialized for the keys involved in K'.

Failure

In the cases where a responsible peer fails, the next responsible peer uses the timestamp values, which are stored along with updates over the members of the group, in order to initialize its counters. Let k be a key whose responsible fails, and p be the peer that is the new responsible for it. For initializing the counter of k, the new responsible peer p contacts the members of the group, retrieves the most recent timestamp which is stored over each member, and selects the highest timestamp as the value of the counter for k, i.e. $c_{p,k}$.

One important question is how the new responsible peer p gets the address of the other members? The answer is as follows. If p was a member of the group before becoming its responsible, it has the address of other members of the group. Thus, it can communicate with them easily. If it is a new member, *i.e.* it has just joined the DHT, it waits until being contacted by the other members of the group. Recall that each member of the group, e.g. q, periodically sends an alive message to the group's responsible peer. If q receives no acknowledge from the responsible peer, it understands that probably the responsible peer has failed. Thus, it uses the lookup service of the DHT to find the address of the peer that is the responsible for $h_r(k)$ in the DHT. If the address is different from the previous one, q gets sure that the responsible peer has changed, so it sends it a message that involves the address of other members of the group. It also contacts the other members of the group to inform them about the modification in the responsibility of the group.

5   Update with Continuous Timestamps

To update the replicated data in the replica holder groups, CTRM uses a new protocol called UCT (Update with Continuous Timestamps). In this section, we describe the details of UCT.

5.1   UCT Protocol

To simplify the description of our UCT protocol, we assume the existence of (not perfect) failure detectors [7] that can be implemented as follows. When we setup a failure detector on a peer p to monitor peer q, the failure detector periodically sends ping messages to q in order to test whether q is still alive (and connected). If the failure detector receives no response from q, then it considers q as a failed peer, and triggers an error message to inform p about this failure.

---

1. On update requester:
    - Send $\{k, pch\}$ to $rsp(k, h_r)$
    - Monitor $rsp(k, h_r)$ using a failure detector
    - Go to Step 8 if $rsp(k, h_r)$ fails

2. On $rsp(k, h_r)$: upon receiving $\{k, pch\}$
    - Set $c_k = c_k + 1$; // increase counter by one
      // initially we have $c_k=0$;
    - Let $ts = c_k$, send $\{k, pch, ts\}$ to other replica holders;
    - Set a timer on, called ackTimer, to a default time

3. On each replica holder: upon receiving $\{k, pch, ts\}$
    - Maintain $\{k, pch, ts\}$ in a temporary memory on disk;
    - Send ack to $rsp(k, h_r)$;

4. On $rsp(k, h_r)$: upon expiring ackTimer
    - If (number of received acks $\geq$ threshold δ) then send "commit" message to the replica holders;
    - Else set $c_k = c_k - 1$, and send "abort" message to the update requester;

5. On each replica holder: upon receiving "commit"
    - Maintain $\{pch, ts\}$ as a committed patch for k.
    - Update the local replica using pch;
    - Send "terminate" message to $rsp(k, h_r)$

6. On $rsp(k, h_r)$: upon receiving the first 'terminate' message
    - Send "terminate" to update requester

7. On update requester: receiving the 'terminate' from $rsp(k, h_r)$
    - Commit the update operation

8. On update requester: upon detecting a failure on $rsp(k, h_r)$
    - If the 'terminate' message is received then commit the update operation;
    - Else, check replica holders, if at least one of them received the 'commit' message then commit the update operation;
    - Else, abort the update operation;

Fig. 1. UCT protocol

Let us now describe the UCT protocol. Let p_0 be the peer that wants to update a data whose ID is k. The peer p_0 is called update requester. Let pch be the patch of the update performed by p_0. Let p_1 be the responsible for the replica holder group of k, i.e. $p_1 = rsp(k, h_r)$. The protocol proceeds as follows (see Figure 3):

- **Update request.** In this phase, the update requester, i.e. p_0, obtains the address of the group's responsible peer, i.e. p_1, by using the DHT's lookup service, and sends to it an update request containing the pair (k, pch). Then, p_0 waits for a commit message from p_1. It also uses a failure detector and monitors p_1. The wait time is limited by a default value, e.g. by using a timer. If p_0 receives the terminate message from p_1, then it commits the operation. If the timer expires or the failure detector reports a fault of p_1, then p_0 checks whether the update has been done or not, i.e. by checking the data at replica holders. If the answer is positive, then the operation is committed, else it is aborted.

- **Timestamp generation and replica publication.** After receiving the update request, p_1 generates a timestamp for k, e.g. ts, by increasing a local counter that it keeps for k, say c_k. Then, it sends (k, pch, ts) to the replica holders, i.e. the members of its group, and asks them to return an acknowledgement. When a replica holder receives (k, pch, ts), it returns the acknowledgement to p_1 and maintains the data in a temporary memory on disk. The patch is not considered as an update before receiving a commit message from p_1. If the number of received acknowledgements is more than or equal to a threshold δ, then p_1 starts the update confirmation phase. Otherwise p_1 sends an abort message to p_0. The threshold δ is a system parameter, e.g. it is chosen in such a way that the probability that δ peers of the group simultaneously fail is almost zero.

- **Update confirmation.** In this phase, p_1 sends the commit message to the replica holders. When a replica holder receives the commit message, it labels $\{pch, ts\}$ as a committed patch for k. Then, it executes the patch on its local replica, and sends a terminate message to p_1. After receiving the first terminate message from replica holders, p_1 sends a terminate message to p_0. If a replica holder does not receive the commit message for a patch, it discards the patch upon receiving a new patch containing the same or greater timestamp value.

Notice that the goal of our protocol is not to provide eager replication, but to have at least δ replica holders that receive the patch and its timestamp. If this goal is attained, the update operation is committed. Otherwise it is aborted, and the update requester should try its update later.

Let us now consider the case of concurrent updates, e.g. two or more peers want to update a data d at the same time. In this case, the concurrent peers send their request to the peer that is responsible for the d's group, say p_1. The peer p_1 determines an order for the requests, e.g. depending on their arrival time or on the distance of requesters if the requests arrive at the same time. Then it processes the requests one by one according their order, i.e. it commits or aborts one request and starts the next one. Thus, concurrent updates make no problem of inconsistency for our replication management service.

5.2 Fault Tolerance of UCT Protocol

Let us now study the effect of peer failures on the UCT protocol and discuss how they are handled. By peer failures, we mean the situations where a peer crashes or gets disconnected from the network abnormally, e.g. without informing the responsible peer. We show that these failures do not block our update protocol. We also show that even in the presence of these failures, the protocol guarantees continuous timestamping, *i.e.* when an update is committed, the timestamp of its patch is only one unit greater than that of the previous one. For this, it is sufficient to show that if the group's responsible peer fails, each generated timestamp is attached with a committed patch, or is aborted. By aborting a timestamp, we mean returning the counter's value to its value before the update operation. During the UCT protocol execution, a failure on the group's responsible peer may happen in one of the following time intervals:

- I_1: **after receiving the update request and before generating the timestamp.** If the group's responsible peer fails in this interval, then after some time, the failure detector detects the failure or the timer timeouts. Afterwards, the update requester checks the update at replica holders, and since it has not been done, the operation is aborted. Therefore, a failure in this interval does not block the protocol, and continuous timestamping is assured because no update is performed.

- I_2: **after I_1 and before sending the patch to replica holders.** In this interval, like in the previous one, the failure detector detects the failure or the timer timeouts, and thus the operation is aborted. The timestamp *ts*, which is generated by the failed responsible peer, is aborted as follows. When the responsible peer fails, its counters get invalid, and the next responsible peer initializes its counter using the greatest timestamp of the committed patches at replica holders. Thus, the counter returns to its value before the update operation. Therefore, in the case of crash in this interval, continuous timestamping is assured.

- I_3: **after I_2 and before sending the commit message to replica holders.** If the responsible peer fails in this interval, since the replica holders have not received the commit, they do not consider their received data as a valid replica. Thus, when the update requester checks the update, they answer that the update has not been done and the operation gets aborted. Therefore, in this case, continuous timestamping is not violated.

- I_4: **after I_3 and before sending the terminate message to the update requester.** In this case, after detecting the failure or timeout, the update requester checks the status of the update in the DHT and finds out that the update has been done, thus it commits the operation. In this case, the update is done with a timestamp which is one unit greater than that of the previous update, thus the property of continuous timestamping is enforced.

6 Network Cost Analysis

In this section, we give a thorough analysis of CTRM's communication cost for both replica retrieval and update operations, and compare them with those of the same operations in two baseline services. Since usually the communicated messages are relatively small, we measure the communication cost in terms of the number of messages.

6.1 Replica Retrieval Cost

In section 3.2, we described the CTRM's operation for retrieving an up-to-date replica. In this section, we give a probabilistic analysis of this operation' cost in terms of the number of messages which should be communicated over the network.

The communication cost of replica retrieval in CTRM consists of the followings: 1) the cost of finding the group's responsible peer, denoted by c_g; 2) the cost of finding the first up-to-date replica at replica holders, denoted by c_{rh}; 3) the cost of returning the replica to the requester, denoted as c_{rt}. The first cost, i.e. c_g, consists of a lookup in the DHT which usually is done in $O(log\ n)$ messages where n is the number of peers of the DHT. For simplicity, we assume that the cost of a lookup is $log\ n$ messages.

The third cost, i.e. c_{rt}, takes simply one message because the first replica holder that maintains an up-to-date replica sends it directly to the requester.

The second cost, i.e. c_{rt}, depends on the number of replica holders which should be contacted for finding the first up-to-date replica. Let n_{rh} denotes the number of replica holders which should be contacted, then $c_{rt} = 2 \times n_{rh}$, i.e. for each replica holder we need one message to contact it and one message as answer. Thus, the total cost of retrieving an up-to-date replica by CTRM is $c_{ctrm} = (log\ n) + 2 \times n_{rh} + 1$. This cost depends on the number of peers in the system, i.e. n, and the number of replica holders which should be contacted by the group's responsible peer, i.e. n_{rh}.

Let us now give a probabilistic approximation of n_{rh}. Let p_{av} be the probability that a replica holder, which is contacted by the responsible peer, maintains an up-to-date replica. In other words, p_{av} is the ratio of the up-to-date replicas over the total number of replica holders, i.e. $/G_k/$. We give a formula for computing the expected value of the number of replica holders which should be contacted for finding the first up-to-date replica, in terms of p_{av} and $/G_k/$. Let X be a random variable which represents the number of replica holders which should be contacted. We have $Prob(X=i) = p_{av} \times (1-p_{av})^{i-1}$, i.e. the probability of having $X=i$ is equal to the probability that $i-1$ first replica holders do not maintain an up-to-date replica and the ith replica holder maintains an up-to-date one. The expected value of X is computed as follows:

$$E(X) = \sum_{i=0}^{|G_k|} i * \Pr ob(X = i)$$

$$E(X) = p_{av} * (\sum_{i=0}^{|G_k|} i * (1 - p_{av})^{i-1}) \tag{1}$$

Equation 1 expresses the expected value of the number of contacted replica holders in terms of p_{av} and $/G_k/$. Thus, we have the following upper bound for $E(X)$ which is solely in terms of p_{av}:

$$E(X) < p_{av} * (\sum_{i=0}^{\infty} i * (1 - p_{av})^{i-1}) \tag{2}$$

Because $p_{av} \leq 1$, by using the theory of series [3], we have the following equation:

$$\sum_{i=0}^{\infty} i * (1 - p_{av})^{i-1} = \frac{1}{(1 - (1 - p_{av}))^2} \qquad (3)$$

Using Equations 3 and 2, we obtain:

$$E(X) < \frac{1}{p_{av}} \qquad (4)$$

The above equation shows that *the expected value of the number of replica holders which should be contacted by the group's responsible peer is less than the inverse of the probability that a replica at a replica holder is up-to-date.*

Example. Assume that at retrieval time *50 %* of the replica holders have an up-to-date replica, *i.e.* $p_{av}=0.5$. Then the expected value for the number of replica holders to be contacted is less than 2, i.e. $n_{rh} \leq 2$. Thus, we have $c_{ctrm} \leq (log\ n) + 5$. In other words, in this example the total cost of replica retrieval in CTRM, i.e. c_{ctrm}, is close to the cost of doing one lookup in the DHT.

6.2 Data Update Cost

Let us now analyze the communication cost of the CTRM's for updating a data using the UCT protocol (described in Section 5.1) in terms of the number of messages. The communication cost consists of the followings: 1) the cost of finding the peer that is responsible for the group, denoted by c_g; 2) the cost of sending the patch to the replica holders, and receiving the acknowledges, denoted as c_{rh}; 4) the cost of committing or aborting the update operation, c_{cm}.

The first cost as shown in Section 6.1, is equal to doing a lookup in the DHT, thus can be estimated as *log n* where *n* is the number of peers in the DHT. Let *r* be the number of replica holders, i.e. $r=/G_k/$. Then, the second cost, i.e. c_{rh}, is at most $r \times 2$. The third cost consists of sending a message to the replica holders and the requester, thus we have $c_{cm} = r + 1$. Thus, in total the communication cost of the update operation by CTRM is *log n + 3×r + 1* where *r* is the number of replicas and *n* the number of peers in the DHT.

6.3 Comparison with Baseline Services

Let us now compare the communication cost of CTRM with that of two baseline services. Although they cannot provide the same functionality as CTRM, the closest prior works to CTRM are the BRICKS project [13], denoted as BRK, and the Update Management Service (UMS) [1]. The assumptions made by these two works are close to ours, e.g. they do not assume the existence of powerful peers. BRK stores the data in the DHT using multiple keys, which are correlated to the data key. To find an up-to-date replica, BRK has to retrieve all replicas. UMS uses a set of *m* hash functions and replicates the data randomly at *m* different peers. To find an up-to-date replica, UMS has to make several lookups in the DHT.

Let *r* be the number of replicas and *n* the number of peers in the DHT. For updating a data, BRK and UMS perform *r* lookups in the DHT. Thus, the cost of update operation in these two services is *O(r × log n)*. By comparing this cost with that of CTRM, i.e. *O((log n) + r)* we see that the update operation in CTRM has a communication cost that is much lower than that of UMS and CTRM.

For retrieving an up-to-date replica, BRK has to retrieve all replicas. The cost of data retrieval in BRK is O $(r \times log\ n)$. The cost of data retrieval in UMS is $O(n_{cu} \times log\ n)$ where c_{cu} is the number of replicas which should be retrieved by using hash functions in order to find an up-to-date replica by UMS. In general, the value of n_{cu} is similar to the value of n_{rh} in CTRM. As we showed previously the communication cost of data retrieval by CTRM is $O(log\ n + n_{rh})$ which is lower than those of both BRK and UMS.

7 Experimental Validation

In this section, we evaluate the performance of CTRM through experimentation over a 64-node cluster and simulation. The experimentation over the cluster was useful to validate our algorithm and calibrate our simulator. The simulation allows us to study scale up to high numbers of peers (up to 10,000 peers).

7.1 Experimental and Simulation Setup

Our experimentation is based on an implementation of the Chord [21] protocol. We tested our algorithms over a cluster of 64 nodes connected by a 1-Gbps network. Each node has two Intel Xeon 2.4 GHz processors, and runs the Linux operating system. To study the scalability of CTRM far beyond 64 peers, we also implemented a simulator using SimJava. After calibration of the simulator, we obtained simulation results similar to the implementation results up to 64 peers.

Our default settings for different experimental parameters are as follows. The latency between any two peers is a random number with normal distribution and a mean of 100 ms. The bandwidth between peers is also a random number with normal distribution and a mean of 56 Kbps (as in [1]). The simulator allows us to perform tests with up to 10,000 peers, after which simulation data no longer fit in RAM and makes our tests difficult. Therefore, the default number of peers is set to 10,000.

In our experiments, we consider a dynamic P2P system, *i.e.* there are peers that leave or join the system. Peer departures are timed by a random Poisson process (as in [18]). The average rate, *i.e.* λ, for events of the Poisson process is $\lambda=1$/second. At each event, we select a peer to depart uniformly at random. Each time a peer goes away, another joins, thus keeping the total number of peers constant (as in [18]).

We also consider peer failures. Let *fail rate* be a parameter that denotes the percentage of peers that leave the system due to a fail. When a peer departure event occurs, our simulator should decide on the type of this departure, i.e. normal leave or fail. For this, it generates a random number which is uniformly distributed in [0..100]; if the number is greater than *fail rate* then the peer departure is considered as a normal leave, else as a fail. In our tests, the default setting for *fail rate* is 5% (as in [1]). In our tests, unless otherwise specified, the number of replicas of each data is 10.

In our tests, we compared our CTRM service with the replication management services in the BRICKS project [13], denoted as BRK, and the Update Management Service (UMS) [1]. Although they cannot provide the same functionality as CTRM (see Section 8), they are closest prior works to CTRM since their assumptions about the P2P system are similar to ours (as explained in Section 6.3).

7.2 Update Cost

Let us first investigate the performance of CTRM's update protocol. We measure the performance of data update in terms of response time and communication cost. By update response time, we mean the time needed to send the patch of an update operation to the peers that maintain the replicas. By update communication cost, we mean the number of messages needed to update a data.

Using our simulator, we ran experiments to study how the response time increases with the addition of peers. Using the simulator, Figure 4 depicts the total number of messages while increasing the number of peers up to 10,000, with the other simulation parameters set as defaults described in Section 7.1. In all three services, the communication cost increases logarithmically with the number of peers. However, the communication cost of CTRM is much better than that of UMS and BRK. The reason is that UMS and BRK perform multiple lookups in the DHT, but CTRM does only one lookup, i.e. only for finding the responsible peer. Notice that each lookup needs $O(log\ n)$ messages where n is the number of peers of the DHT.

Figure 5 shows the update response time with the addition of peers up to 10,000, with the other parameters set as described in Section 7.1. The response time of CTRM is a little bit higher than that of UMS and BRK. The reason is that for guaranteeing continuous timestamping, the update protocol of CTRM performs two round-tips between the responsible peer and the other members of the group. But UMS and BRK only send the update actions to the replica holders by looking up the replica holders in parallel (note that the impact of parallel lookups on response time is very slight, but they have a high impact on communication cost). However, the difference in the response time of CTRM and that of UMS and BRK is small because the round-trips in the group are less time consuming than lookups. This slight increase in response time of CTRM's update operation is the price to pay for guaranteeing continuous timestamping.

7.3 Data Retrieval Response Time

We now investigate the data retrieval response time of CTRM. By data retrieval response time, we mean the time to return an up-to-date replica to the user.

Figure 6 shows the response time of CTRM, UMS and BRK with the addition of peers up to 10,000, with the other parameters set as defaults described in Section 7.1. The response time of CTRM is much better than that of UMS and BRK. This difference in response time can be explained as follows. Both CTRM and UMS services contact some replica holders, say r, in order to find an up-to-date replica, e.g.
$r=6$. For contacting these replica holders, CTRM performs only one lookup (to find the group's responsible peer) and some low-cost communications in the group. But, UMS performs exactly r lookups in the DHT. BRK retrieves all replicas of data from the DHT (to determine the latest version), and for each replica it performs one lookup. Thus the number of lookups done by BRK is equal to the total number of data replicas, i.e. 10 in our experiments.

Let us now study the effect of the number of replicas of each data, say m, on performance of data retrieval. Figure 7 shows the response time of data retrieval for the three solutions while varying the number of replicas up to 30. The number of replicas has almost a linear impact on the response time of BRK, because to retrieve an up-to-date replica it has to retrieve all replicas by doing one lookup for each replica. But it has a slight impact on CTRM, because for finding an up-to-date replica CTRM performs only one lookup, and some low cost communications, i.e. in the group.

7.4 Effect of Peer Failures on Timestamp Continuity

Let us now study the effect of peer failures on the continuity of timestamps used for data updates. This study is done only for CTRM and UMS that work based on time-stamping. In our experiments we measure *timestamp continuity rate* by which we mean the percentage of the updates whose timestamps are only one unit higher than

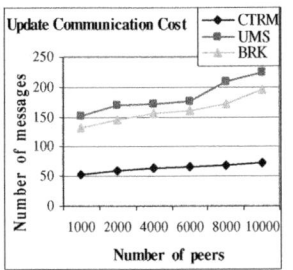
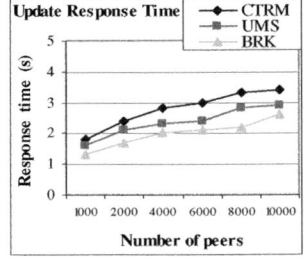
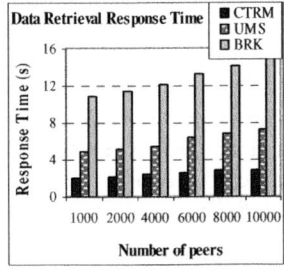

Fig. 4. Communication cost of updates vs. number of peers

Fig. 5. Response time of update operation vs. number of peers

Fig. 6. Response time of data retrievals vs. number of peers

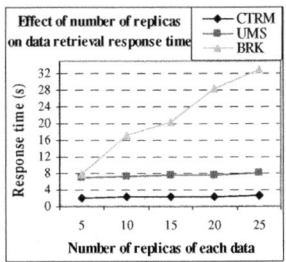
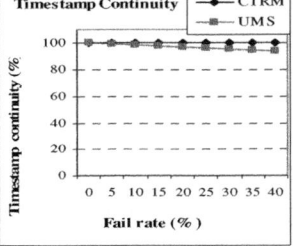
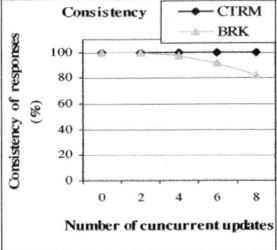

Fig. 7. Effect of the number of replicas on response time of data retrievals

Fig. 8. Timestamp continuity vs. fail rate

Fig. 9. Consistency of returned results vs. number of concurrent updates

that of their precedent update. We varied the fail rate parameter, and observed its effect on timestamp continuity rate.

Figure 8 shows timestamp continuity rate for CTRM and UMS while increasing the fail rate, with the other parameters set as described in Section 7.1. The peer failures do not have any negative impact on the continuity of timestamps generated by CTRM, because our protocol assures timestamp continuity. However, when increasing the fail rate in UMS, the percentage of updates whose timestamps are not continuous increases.

7.5 Effect of Concurrent Updates on Result Consistency

In this section, we investigate the effect of concurrent updates on the consistency of the results returned by CTRM. In our experiments, we perform u updates done concurrently by u different peers using the CTRM service, and after finishing the concurrent updates, we invoke the service's data retrieval operation from n randomly chosen peers ($n=50$ in our experiments). If there is any difference between the data returned to the n peers, we consider the result as inconsistent. We repeat each experiment several times, and report the percentage of the experiments where the results are consistent. We perform the same experiments using the BRK service.

Figure 9 shows the results with the number of concurrent updates, i.e. u, increasing up to 8, and with the other parameters set as defaults described in Section 7.1. As shown, in 100% of experiments the results returned by CTRM are consistent. This shows that our update protocol works correctly even in the presence of concurrent updates. However, the BRK service cannot guarantee the consistency of results in the case of concurrent updates, because two different updates may have the same version at different replica holders.

8 Related Work

In the context of distributed systems, data replication has been widely studied to improve both performance and availability. Many solutions have been proposed in the context of distributed database systems for managing replica consistency [4][15], in particular, using eager or lazy (multi-master) replication techniques, e.g. [12][14][24]. However, these techniques either do not scale up to large numbers of peers or raise open problems, such as replica reconciliation, to deal with the open and dynamic nature of P2P systems.

Most existing P2P systems support data replication, but without consistency guarantees. For instance, Gnutella [9] and KaZaA [11], two of the most popular P2P file sharing systems allow files to be replicated. However, a file update is not propagated to the other replicas. As a result, multiple inconsistent replicas under the same identifier (filename) may co-exist and it depends on the peer that a user contacts whether a current replica is accessed. In Freenet [6], the query answers are replicated along the path between the peers owning the data and the query originator. In the case of an update (which can only be done by the data's owner), it is routed to the peers having a replica. However, there is no guarantee that all those peers receive the update, in particular those that are absent at update time.

PGrid is a structured P2P system that deals with data replication and update based on a gossiping algorithm [8]. It provides a fully decentralized update scheme, which offers probabilistic guarantees. However, replicas may get inconsistent, *e.g.* as a result of concurrent updates, and it is up to the users to cope with the problem.

OceanStore [19] is a data management system designed to provide a highly available storage utility on top of P2P systems. It allows concurrent updates on replicated data, and relies on reconciliation to assure data consistency. The reconciliation is done by a set of high performance nodes, using a consensus algorithm. These nodes agree on which operations to apply, and in what order. However, in the applications that we address, the presence of such nodes is not guaranteed.

In [10], the authors present a performance evaluation of different strategies for placing the data replicas in DHTs. Our solution can be classified among those that use the neighbor replication strategy, i.e. that tries to place the replicas over the neighbors of a peer. However, our update protocol is new and not covered by any of the strategies described in [10].

The BRICKS project [13] provides high data availability in DHTs through replication. For replicating a data, BRICKS stores the data in the DHT using multiple keys, which are correlated to the data key, e.g. k. There is a function that, given k, determines its correlated keys. To be able to retrieve an up-to-date replica, BRICKS uses versioning. Each replica has a version number which is increased after each update. However, because of concurrent updates, it may happen that two different replicas have the same version number, thus making it impossible to decide which one is the latest replica.

In [1], an update management service, called UMS, was proposed to support data currency in DHTs, *i.e.* the ability to return an up-to-date replica. However, UMS does not guarantee continuous timestamping which is a main requirement for collaborative applications which need to reconcile replica updates. UMS uses a set of m hash functions and replicates randomly the data at m different peers, and this is more expensive than the groups which we use in CTRM, particularly in terms of communication cost. A prototype based on UMS was demonstrated in [22].

9 Conclusion

In this paper, we addressed the problem of efficient replication management in DHTs. We proposed a new service, called continuous timestamp based replication management (CTRM), which deals with efficient data replication, retrieval and update in DHTS, by taking advantage of replica holder groups which are managed dynamically. We dealt with the dynamic behaviour of the group members, which can leave the system at any time. To perform updates on replicas, we proposed a new protocol that stamps the updates with timestamps that are generated using the replica holder groups. The updates' timestamps are not only monotonically increasing but also continuous. We analyzed the communication cost of CTRM, and show that its cost is very low in comparison to two baseline services in DHTs.

We evaluated CTRM through experimentation and simulation; the results show its effectiveness for data replication in DHTs. The results of our evaluation show that with a low overhead in update response time, CTRM supports fault-tolerant data replication using continuous timestamps. In our experiments, we compared CTRM with two baseline services, and the results show that data retrieval with CTRM is much more efficient than the baseline services. We investigated the effect of peer failures on the correctness of CTRM and the results show that it works correctly even in the presence of peer failures.

References

[1] Akbarinia, R., Pacitti, E., Valduriez, P.: Data Currency in Replicated DHTs. In: ACM Int. Conf. on Management of Data (SIGMOD), pp. 211–222 (2007)

[2] Akbarinia, R., Tlili, M., Pacitti, E., Valduriez, P., Lima, A.A.B.: Continuous timestamping for efficient replication management in dHTs. In: Hameurlain, A., Morvan, F., Tjoa, A.M. (eds.) Globe 2010. LNCS, vol. 6265, pp. 38–49. Springer, Heidelberg (2010)

[3] Bromwich, T.J.I.: An Introduction to the Theory of Infinite Series, 3rd edn. Chelsea Pub. Co., New York (1991)

[4] Cecchet, E., Candea, G., Ailamaki, A.: Middleware-based database replication: the gaps between theory and practice. In: ACM Int. Conf. on Management of Data (SIGMOD), pp. 739–752 (2008)

[5] Chawathe, Y., Ramabhadran, S., Ratnasamy, S., LaMarca, A., Shenker, S., Hellerstein, J.M.: A case study in building layered DHT applications. In: ACM Conf. on Applications, Technologies, Architectures, and Protocols for Computer Communication (SIGCOMM), pp. 97–108 (2005)

[6] Clarke, I., Miller, S.G., Hong, T.W., Sandberg, O., Wiley, B.: Protecting Free Expression Online with Freenet. IEEE Internet Computing 6(1), 40–49 (2002)

[7] Dabek, F., Kaashoek, M.F., Karger, D., Morris, R., Stoica, I.: Wide-Area Cooperative Storage with CFS. In: ACM Symp. on Operating Systems Principles, pp. 202–215 (2001)

[8] Datta, A., Hauswirth, M., Aberer, K.: Updates in Highly Unreliable, Replicated Peer-to-Peer Systems. In: IEEE Int. Conf. on Distributed Computing Systems (ICDCS), pp. 76–87 (2003)

[9] Gnutella, http://www.gnutelliums.com/

[10] Ktari, S., Zoubert, M., Hecker, A., Labiod, H.: Performance evaluation of replication strategies in DHTs under churn. In: Int. Conf. on Mobile and Ubiquitous Multimedia (MUM), pp. 90–97 (2007)

[11] Kazaa, http://www.kazaa.com/

[12] Krikellas, K., Elnikety, S., Vagena, Z., Hodson, O.: Strongly consistent replication for a bargain. In: IEEE Int. Conf. on Data Engineering (ICDE), pp. 52–63 (2010)

[13] Knezevic, P., Wombacher, A., Risse, T.: Enabling High Data Availability in a DHT. In: Proc. of Int. Workshop on Grid and P2P Computing Impacts on Large Scale Heterogeneous Distributed Database Systems, pp. 363–367 (2005)

[14] Lin, Y., Kemme, B., Jiménez-Peris, R., Patiño-Martínez, M., Armendáriz-Iñigo, J.E.: Snapshot isolation and integrity constraints in replicated databases. ACM Transactions on Database Systems (TODS) 34(2) (2009)

[15] Özsu, T., Valduriez, P.: Principles of Distributed Database Systems, 2nd edn. Prentice-Hall, Englewood Cliffs (1999)

[16] PalChaudhuri, S., Saha, A.K., Johnson, D.B.: Adaptive Clock Synchronization in Sensor Networks. In: Int. Symp. on Information Processing in Sensor Networks, pp. 340–348 (2004)

[17] Ratnasamy, S., Francis, P., Handley, M., Karp, R., Shenker, S.: A scalable content-addressable network. In: ACM Conf. on Applications, Technologies, Architectures, and Protocols for Computer Communication (SIGCOMM), pp. 161–172 (2001)

[18] Rhea, S.C., Geels, D., Roscoe, T., Kubiatowicz, J.: Handling churn in a DHT. In: USENIX Annual Technical Conf, pp. 127–140 (2004)

[19] Rhea, S.C., Eaton, P., Geels, D., Weatherspoon, H., Zhao, B., Kubiatowicz, J.: Pond: the OceanStore Prototype. In: USENIX Conf. on File and Storage Technologies, pp. 1–14 (2003)

[20] Rowstron, A., Druschel, P.: Pastry: Scalable, decentralized object location, and routing for large-scale peer-to-peer systems. In: Liu, H. (ed.) Middleware 2001. LNCS, vol. 2218, pp. 329–350. Springer, Heidelberg (2001)

[21] Stoica, I., Morris, R., Karger, D.R., Kaashoek, M.F., Balakrishnan, H.: Chord: a scalable peer-to-peer lookup service for internet applications. In: ACM Conf. on Applications, Technologies, Architectures, and Protocols for Computer Communication (SIGCOMM), pp. 149–160 (2001)

[22] Tlili, M., Dedzoe, W.K., Pacitti, E., Valduriez, P., Akbarinia, R., Molli, P., Canals, G., Laurière, S.: P2P logging and timestamping for reconciliation. PVLDB 1(2), 1420–1423 (2008)

[23] Xwiki Concerto Project, http://concerto.xwiki.com

[24] Wong, L., Arora, N.S., Gao, L., Hoang, T., Wu, J.: Oracle Streams: A High Performance Implementation for Near Real Time Asynchronous Replication. In: IEEE Int. Conf. on Data Engineering (ICDE), pp. 1363–1374 (2009)

A Survey of Structured P2P Systems for RDF Data Storage and Retrieval*

Imen Filali, Francesco Bongiovanni, Fabrice Huet, and Françoise Baude

INRIA Sophia Antipolis
CNRS, I3S, University of Nice Sophia Antipolis
France
first.last@inria.fr

Abstract. The Semantic Web enables the possibility to model, create and query resources found on the Web. Enabling the full potential of its technologies at the Internet level requires infrastructures that can cope with scalability challenges while supporting expressive queries. The attractive features of the Peer-to-Peer (P2P) communication model, and more specifically *structured P2P systems*, such as decentralization, scalability, fault-tolerance seems to be a natural solution to deal with these challenges. Consequently, the combination of the Semantic Web and the P2P model can be a highly innovative attempt to harness the strengths of both technologies and come up with a scalable infrastructure for RDF data storage and retrieval. In this respect, this survey details the research works adopting this combination and gives an insight on how to deal with the RDF data at the indexing and querying levels. We also present some works which adopt the publish/subscribe paradigm for processing RDF data in order to offer long standing queries.

Keywords: Semantic Web, Peer-to-Peer (P2P), Distributed Hash Tables (DHTs), Resource Description Framework (RDF), Distributed RDF repository, RDF data indexing, RDF query processing, publish/subscribe (pub/sub), subscription processing.

1 Introduction

The realization of the Semantic Web [7] and more specifically the "Linked Data" vision [18] of Sir Tim Berners Lee at the Internet level has seen the emergence of a new breed of distributed systems that combines Peer-to-Peer (P2P) technologies with the Resource Description Framework (RDF) (metadata) data model [4]. This combination, which allows a flexible description and sharing of both data and metadata in large scale settings, arose quite naturally, and, as stated in [82], both address the same need but at different levels. The Semantic Web allows

* The presented work is funded by the EU FP7 NESSI strategic integrated project SOA4All (http://www.soa4all.eu), EU FP7 STREP project PLAY (http://www.play-project.eu) and French ANR project SocEDA (http://www.soceda.org)

A. Hameurlain, J. Küng, and R. Wagner (Eds.): TLDKS III , LNCS 6790, pp. 20–55, 2011.

users to model, manipulate and query knowledge at a conceptual level with the intent to exchange it. P2P technologies, on the other hand, enable users to share the information, encompassing the need of its indexing and storage, using a decentralized organization. P2P systems have been recognized as a key communication model to build scalable platforms for distributed applications such as file sharing (e.g., Bittorrent [29]) or distributed computing (e.g., SETI@home [5]). P2P architectures are classified into three main categories: *unstructured, hybrid* and *structured* overlays. In *unstructured* P2P overlays, there is no constraint on the P2P architecture as the overlay is built by establishing random links among nodes [1]. Despite their simplicity, they suffer from several issues: limited scalability, longer search time, higher maintenance costs, etc. A *hybrid* P2P overlay is an architecture in which some parts are built using random links and some others, generally the core of the overlay, are constructed using a specific topology [87]. *Structured Overlay Networks* (SONs) [36] emerged to alleviate inherent problems of unstructured overlays. In these systems, peers are organized in a well-defined geometric topology (e.g., a ring, a torus, etc.) and, compared to unstructured networks, they exhibit stronger guarantees in terms of search time and nodes' maintenance [27]. By providing a *Distributed Hash Table (DHT)* abstraction, by the means of *Put* and *Get* methods, they offer simple mechanisms to easily store and fetch the data in a distributed environment. However, even if this abstraction is practical, it has its limits when it comes to *efficiently* support the types of queries that the Semantic Web technologies intend to provide. The main goal behind these systems is a simple *data storage* and *retrieval*. Therefore, a key point in building a scalable distributed system is how to efficiently index data among peers participating in the P2P network in order to ensure efficient lookup services, and thus improve the efficiency of applications and services executed at the top level.

The question we try to address in this survey is the following: how to store the RDF data and evaluate complex queries expressed in various Semantic Web query languages (e.g., [6, 50, 81]) on top of fully distributed environments? In this regard, we survey several approaches for P2P distributed storage systems that adopt RDF as a data model. In particular, we focus on data indexing, data lookup and query evaluation mechanisms. We also give an overview of the publish/subscribe (pub/sub) communication paradigm build on top of P2P systems for RDF data storage and retrieval. Rather than pursuing an exhaustive list of P2P systems for RDF data storage and retrieval, our aim is to provide a unified view of algorithmic approaches in terms of data indexing and retrieval, so that different systems can be put in perspective, compared and evaluated. Several survey papers focusing on P2P networks (e.g., [51, 62, 68, 77]) or the pub/sub communication model (e.g., [38, 60]) have been proposed. None of them addressed the combination of the P2P architecture and the RDF data model. The work presented by Lua *et al.* in [62] gives an extensive comparative study of the basic structured and unstructured P2P systems which is out of the scope of this paper. Related surveys have been presented in [68, 77] but even if they have a wider scope than [62], as they cover several service discovery frameworks, search methods, etc., they do not detail the proposed mechanisms. This survey, on the contrary, presents an in depth analysis of carefully selected papers that combine the RDF data model and the P2P technologies.

In order to be self-contained, Section 2 introduces the context of this paper and gives an insight of the basic terminologies and technologies that will be considered while investigating RDF-based P2P systems. This includes the main concepts behind the RDF data model as well as a taxonomy of P2P systems from the overlay architecture point of view. Section 3 discusses several P2P systems that combine a P2P architecture with the RDF data model. Specifically, it focuses on the data indexing and query processing mechanisms and also investigates the pub/sub paradigm on top of RDF-based P2P systems. Moreover, it underlines various algorithms that have been proposed to manage complex subscriptions. Summary and general discussions are laid out in Section 4 before concluding the survey in Section 5.

2   Context and Background

The main idea behind the Semantic Web is to define and link the Web content in a machine understandable way. Realizing this vision requires well defined knowledge representation models and languages to create, model and query resources on the Web. Several technologies have been proposed in this direction to accomplish this goal at the conceptual level [7]. The processing of a huge amount of information at the Web scale, from an architectural point of view, is not a trivial task and the ever-growing increase of the resources available on the Web is overwhelming for traditional centralized systems. Client/server-based solutions suffer from scalability issues so there is a need for a (fully) distributed solution to cope with this problem. After the success that has been revealed by the P2P communication model, the idea of exploiting its strengths (e.g., distributed processing, fault-tolerance, scalability, etc.) to answer the requirements of the Semantic Web layers, and more specifically the RDF data model, has been explored. Besides the attempt to combine the strengths of the RDF and the P2P models, the need for *up-to-date* data requires more advanced algorithms and techniques in order to immediately retrieve data when it gets stored in the system. The publish/subscribe communication paradigm seems to be a natural solution to deal with this issue. In this respect, we will investigate along the remainder of this paper to which extent the Semantic Web can benefit from the P2P community in accordance with RDF data storage and retrieval.

Before presenting and discussing a set of selected works,we give in the following section the basic concepts behind the RDF model. Then, we briefly describe the main overlay architectures as well as the key ideas behind the pub/sub paradigm. Finally, we outline some challenges related to RDF data and query processing on top of P2P systems. Note that although several technologies such as the Resource Description Framework Schema (RDFS), the Web Ontology Language (OWL) are among the main building blocks of Semantic Web stack, we limit ourselves to RDF data storage and retrieval and related query mechanisms on top of P2P systems.

2.1   The RDF Data Model

The Resource Description Framework (RDF) [4] is a W3C standard aiming to improve the World Wide Web with machine processable semantic data. RDF provides a powerful abstract data model for knowledge representation which can be serialized in XML or Notation3 (N3) formats. It has emerged as the prevalent data model for the Semantic Web [7] and is used to describe semantic relationships among data. The notion of RDF *triple* is the basic building block of the RDF model. It consists of a *subject (s)*, a *predicate (p)* and an *object (o)*. More precisely, given a set of *URI* references $\mathcal{R}$, a set of blank nodes $\mathcal{B}$, and a set of literals $\mathcal{L}$, a triple $(s, p, o) \in (\mathcal{R} \cup \mathcal{B}) \times \mathcal{R} \times (\mathcal{R} \cup \mathcal{B} \cup \mathcal{L})$ The subject of a triple denotes the *resource* that the statement is about, the predicate denotes a *property* or a characteristic of the subject, and the object presents the *value* of the property. Note that there is another way to look for the RDF information which is a labeled directed connected graph. Nodes are the subjects and the objects of the RDF statements, while edges refer to the predicates of the triples. Edges are always directed from the subjects to the objects. This presentation gives an easy to understand visual explanation of the relationship between the RDF triples elements. Figure 1 shows an example of a RDF graph.

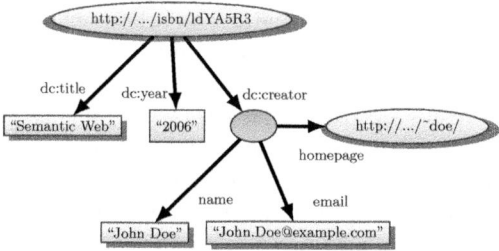

Fig. 1. An Example of a RDF graph. Plain nodes literals are presented with rectangles. Nodes with URI references are presented by oval shapes. Empty gray node presents a blank node.

RDFS [32] is an extension of RDF providing mechanisms for describing groups of resources and their relationships. Among other capabilities, it allows to define classes of resources and predicates. Moreover, it enables to specify constraints on the subject or the object of a given class of predicates.

In addition to the definition of the RDF data model which will be considered in our further discussions, we need to define and classify different types of queries considered in the presented works:

- **Atomic queries** are triple patterns where the subject, predicate and object can either be variables or constant values. According to [23], atomic triple queries can be summarized into eight queries patterns: $(?s, ?p, ?o)$, $(?s, ?p, o_i)$, $(?s, p_i, ?o)$, $(?s, p_i, o_i)$, $(s_i, ?p, ?o)$, $(s_i, ?p, o_i)$, $(s_i, p_i, ?o)$, (s_i, p_i, o_i). In each query pattern, $?x$ denotes the element(s) that the query is looking for and x_i refers to a constant value of the RDF element. For instance, the query pattern $(?s, p_i, o_i)$ returns all subjects s for a given predicate p_i and object o_i (if they exist).

- **Conjunctive queries** are expressed as a conjunction of a list of atomic triple patterns (sub-queries). For instance, a conjunctive query may look like: $(?s_1, p_1, o_1) \wedge (?s_2, p_2, o_2)$.
- **Range queries** involve queries for single or multiple attribute data whose attribute value falls into a range defined by the query. For instance $q = (s, p, ?o)$ FILTER $o < v$, looks, for a given subject s and a predicate p, for all object values bounded by the value v.
- **Continuous queries** are a particular type of queries which are decoupled in time, space and synchronization between peers participating in the query resolution. This kind of querying is found in pub/sub systems where subscribers of a given query can continuously receive matching results in the form of asynchronous notifications. Any type of queries (atomic, conjunctive, etc.) can be used in an asynchronous manner.

2.2 P2P Architectures

Unstructured P2P overlays. In this model, peers form a loosely coupled overlay without any constraint neither on the network topology nor on the data placement. Peers are inherently flexible in their neighbor selection, that is, when a new peer joins the P2P network, it freely selects its neighbors to connect to. The resulting graph characteristics, such as peers' degree, depend on the way peers are connected. Therefore, connections between peers may be based on purely randomized approaches or it can benefit,for instance, from other aspects such as semantic clustering [30]. Despite the unstructured P2P architectures are appealing due to their properties of fault-tolerance, these overlays often provide restricted guarantees to locate the data even if it is available in the network as the life time of the query is generally limited by a fixed number of hops. Techniques based on *gossip-based* algorithms [34, 37], inspired by the way infections diseases spread, offer solutions for improving search in unstructured overlays.

Hybrid P2P overlays. In this model, a set of peers act as super-peer nodes for a set of "ordinary" peers and can perform more complex tasks such as data indexing and query routing (e.g., [61, 72, 87]). Super-peers could be connected to each other in a structured manner while the rest of the nodes establishes random connections with one another.

Structured P2P overlays. Unlike unstructured P2P systems, structured P2P overlays, are built according to a well-defined geometric structure as well as deterministic links between nodes. Due to these properties, they can offer guarantees regarding the lookup time, scalability and can also decrease the maintenance cost of the overlay [27]. P2P overlays such as CAN [75], Chord [83], Pastry [78], P-Grid [8] and many others present an attractive solution for building large scale distributed applications thanks to the practical DHT abstraction that they offer. Although they differ in several design choices such as the network topology, the routing mechanisms, they provide a *lookup* function on top of which they offer this DHT abstraction. Most of DHT-based systems share a common idea, that is, the mapping of a key space to a set of peers such that each peer is responsible for a partition of that space. The *Put* and *Get* methods provided by the DHT abstraction are respectively responsible for storing and retrieving the data. The use of hashing functions by DHTs guarantees, with high probability, a uniform distribution of the data. This valuable property correlated with all previous ones make DHTs the most suitable choice as a P2P substrate.

2.3 Publish/Subscribe in P2P Systems

In this section, we briefly highlight the main concepts underlying the pub/sub communication model. We then focus on the synergy between this particular paradigm and RDF-based P2P systems, in order to see how complementary a dynamic messaging paradigm and a rich data model could be. By going into details of the subscription and advertisement processing algorithms, we will present some ground works that were done around this promising symbiosis.

In a pub/sub system, subscribers, also called consumers, can express their interests in an event or a pattern of events, and be notified of any generated event by the publishers (producers) that matches those interests. The events are propagated asynchronously to all subscribers. Therefore, the overall system is responsible for matching the events to the subscriptions and for the delivery of those relevant events from the publishers to the subscribers which are distributed across a wide area network. This paradigm provides a decoupling in time, space and synchronization between participants. First, subscribers and publishers do not need to participate in the relation at the same time. Secondly, senders and receivers are not required to have a prior knowledge of each other and can even be located in separate domains. Finally, they are not blocked when generating events and subscribers can get the notifications in an asynchronous manner. Overall, the key components of a pub/sub system can be summarized into the following concepts:

– **Subscriptions.** A subscription describes a set of notifications a consumer is interested in. The goal behind the subscription process is that subscribers will receive notifications matching their interests from other peers in the network. Subscriptions are basically *filters*, which can range from simple boolean-valued functions to the use of a complex query language. The expressiveness of the subscriptions in terms of filtering capabilities depends directly from the data and the filter models used. Note that when discussing the pub/sub model, the words "subscription" and "query" are interchangeable.
– **Notifications.** In pub/sub systems, notifications can signify various types of events depending on the perspective. From a publisher perspective, *advertisements* are a type of notification used to describe the kind of notification the publishers are willing to send. From a subscriber perspective, a *notification* is an event that matches the subscription(s) of consumers which is responsible for conveying notifications to subscribers. Several peers within

the network can actively or passively participate in the dissemination of those notifications.

Pub/sub systems have several ways for identifying notifications that can be based either on a *topic* or the *content*. In the *topic-based* model, publishers annotate every event they generate with a string denoting a distinct topic. Generally, a topic is expressed as a rooted path in a tree of subjects. For instance, an online research literature database application (e.g., Springer) could publish notifications of newly available articles from the *Semantic Web* research area under the subject "/Articles/Computer Science/Proceedings/Web/Semantic Web". This kind of topic will then be used by subscribers which will, upon subscription's generation, explicitly specify the topic they are interested in and for which they will receive every related notifications. The topic-based model is at the core of several systems such as Scribe [26], Sub-to-Sub [86]. A main limitation of this model lies in the fact that a subscriber could be interested only in a subset of events related to a given topic instead of all events. This comes from the tree-based classification which severely constrains the expressiveness of the model as it restricts notifications to be organized using a single path in the tree. A tree-based topic hierarchy inhibits the usage of multiple super-topics for instance, even if some inner re-organizations are possible, this classification mechanism remains too rigid. On the other side, a *content-based* model is much more expressive since it allows the evaluation of filters on the whole content of the notifications. In other words, it is the data model and the applied predicates that exclusively determine the expressiveness of the filters. Subscribers express their interests by specifying predicates over the content of notifications they want to receive. These constraints can be more or less complex depending on the query types and operators that are offered by the system. Available query predicates range from simple comparisons, regular expressions, to conjunctions, disjunctions, and XPath expressions on XML. As the focus of this survey is RDF data storage and retrieval on top of P2P networks, we only consider pub/sub systems which solely address such combination. Therefore, this overview complements other surveys related to the pub/sub paradigm. The interested reader can refer to [38, 60, 70] for deeper discussions on the basic concepts behind this communication model.

2.4   RDF Data Processing on Top of P2P Systems: What Are the Main Challenging Aspects ?

The P2P communication model has drawn much attention and has emerged as a powerful architecture to build large scale distributed applications. Earlier research efforts on P2P systems have focused on improving the scalability of unstructured P2P systems by introducing specific geometries for the overlay network, i.e., SONs and adding a standard abstraction on top of it, i.e., DHT. The first wave of SONs [75, 83] focused on scalability, routing performances and churn resilience. The main issue of these DHTs is that they only support *exact queries*, i.e., querying for data items matching a given key (lookup(key)). To overcome such limitation, a second wave of SONs led to a set of P2P approaches such as [8, 22] having the capabilities to manage more complex queries such as range queries [8] or multi-attribute queries [22]. However, storage and retrieval of RDF data in a distributed P2P environment raise new challenges. In essence, as the RDF data model supports more complex queries such as conjunctive and disjunctive queries, an additional effort to design adequate algorithms and techniques to support advanced querying beyond simple keyword-based retrieval and range queries is required. Consequently, a particular attention has to be made regarding the data indexing since it has a great impact on the query processing phase. Overall, two main aspects have to be taken into consideration while investigating RDF data storage and retrieval in P2P systems:

Data indexing. How can we take advantage of the format of the RDF data model in order to efficiently index RDF triples?

Query processing. How can we take advantage of P2P search mechanisms to efficiently query and retrieve RDF data in large scale settings? Moreover, whenever a query, composed of a set of sub-queries, can be answered by several nodes, how to efficiently combine RDF data residing in different locations before sending the result back to the user? In addition, what is the impact of the data indexing methods on the query processing algorithms?

3   RDF Data Storage and Retrieval in P2P Systems

Centralized RDF repositories and lookup systems such as RDFStore [3], Jena [2], RDFDB [79] have been designed to support RDF data storage and retrieval. Although these systems are simple in design, they suffer from the traditional limitations of centralized approaches. As an alternative to these centralized systems, P2P approaches have been proposed to overcome some of these limitations by building (fully) decentralized data storage and retrieval systems. The remainder of this section presents a selection of research works combining the RDF data model with a P2P model with a focus on data indexing and query processing. Note that we do not consider in our discussions unstructured P2P approaches for RDF data storage and retrieval such as Bibster [46] and S-RDF [88], that we have discussed in [39] and we only overview P2P systems based on structured overlays. These systems differ in the overlay architecture, data indexing and query processing mechanisms. These approaches, presented in the next sections, are classified according to the overlay structure (ring-based, cube-based, tree-based) in case they are directly related to the underlying structure. Otherwise, they are grouped into a class called "generic DHT-based". In Section 3.5, we present a set of pub/sub systems dealing with asynchronous queries. Therefore, this overview complements other surveys related to the pub/sub paradigm. The interested reader can refer to [38, 60, 70] for deeper discussions on the basic concepts behind this communication model. Note that along the remainder of this work, we use the original notation found in the presented papers.

3.1   Ring-Based Approaches

RDFPeers. RDFPeers [23] was the first P2P system that has proposed the use of DHTs in order to implement a distributed RDF repository. It is built

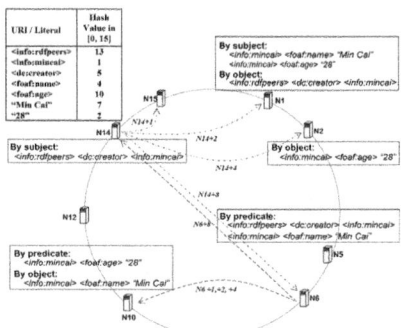

Fig. 2. Example of RDF data storage in RDFPeers network with 8 present nodes in a 4 bit identifier space (taken from[23])

on top of MAAN [22](Multiple Attribute Addressable Network) which is an extension of Chord [83]. As in Chord, each node has an identifier that represents its position in a circular identifier space of size N, and has direct references to its predecessor and successor in the ring. Moreover, a node keeps $\mathtt{m} = \mathtt{log(N)}$ routing entries, called *fingers*. A $\mathtt{finger[i]}$ equals to $\mathtt{successor}(\mathtt{n} \oplus 2^{i-1})$ with $1 \leq \mathtt{i} \leq \mathtt{m}$, where $(\mathtt{n} \oplus 2^{i-1})$ is $(\mathtt{n} + 2^{i-1})$ modulo N. A successor of a node n is defined as the first node that succeeds n along the clockwise direction in the ring space. Figure 2 shows a RDFPeers network with 8 present nodes in a 4 bit identifier space. It also depicts the fingers of nodes N_6 and N_{14} presented by dashed arrows originated respectively from N_6 and N_{14}. Data indexing and query processing mechanisms in RDFPeers work as follows:

Data indexing. RDFPeers builds a distributed index. More precisely, a RDF triple, labeled $\mathtt{t=(s,p,o)}$, is indexed three times by applying a hash function on the subject ($\mathtt{hash(s)}$), the predicate ($\mathtt{hash(p)}$) and the object ($\mathtt{hash(o)}$). The triple t will then be stored on peers responsible for those hashed values. For instance, to store a triple t according to the predicate element, the node which is responsible for $\mathtt{hash(p)}$ will store the triple t. For better understanding consider the storage of the following three triples in the RDFPeers infrastructure depicted in Figure 2: $\mathtt{<info:rdfpeers><dc:creator><info:mincai>}$; $\mathtt{<info:mincai><foaf:name>"Min\ Cai"}$; $\mathtt{<info:mincai>\ <foaf:age>\ "28"\ \hat{\ }\hat{\ }}$ $\mathtt{<xmls:integer>}$. Figure 2 depicts the hashed values of all the URI and literals values of these triples. Suppose now that node N_6 receives a $\mathtt{store}$ message aiming to insert the triple $\mathtt{<info:mincai>\ <foaf:age>\ "28"\ \hat{\ }\hat{\ }<xmls:integer>}$ in the network according the hashing of the predicate element (i.e., $\mathtt{<foaf:age>}$). As $\mathtt{hash(<foaf:age>)=10}$ (see table in Figure 2), N_6, and based on its fingers, will route the query to the N_{10}.

Query processing. RDFPeers supports three kinds of queries.

- **Atomic triple pattern query.** In this kind of query, the subject, the predicate, or the object can either be a variable or an exact value. For instance, a query like $(s,?p,?o)$ will be forwarded to the node responsible for $hash(s)$. All atomic queries take $\mathcal{O}(log(N))$ routing hops to be resolved except queries in the form of $(?s,?p,?o)$ which require $\mathcal{O}(N)$ hops in a network of N peers.

- **Disjunctive and range query.** RDFPeers optimizes this kind of query through the use of the locality preserving hash function. Indeed, when the variable's domain is limited to a range, the query routing process starts from the node responsible for the lower bound. It is then forwarded linearly until received by the peer responsible for the upper bound. In the case of disjunctive range query like $(s, p, ?o)$, $?o \in \cup_{i=1}^n[l_i, u_i]$, such as l_i and u_i are respectively the lower and the upper bound of the range i, several ranges have to be satisfied, intervals are sorted in ascending order. The query will then be forwarded from one node to the other, until it will be received by the peer responsible for the upper bound of the last range. Disjunctive exact queries such as $(s, p, ?o)$, $o? \in \{v_1, v_2\}$ are resolved using the previous algorithm since they are considered as a special case of disjunctive range queries where the lower and the upper bounds are equal to the exact match value.

- **Conjunctive query.** RDFPeers supports this type of queries, as long as they are expressed as a conjunction of atomic triples patterns or disjunctive range queries for the *same* subject. Constraints' list can be related to predicates or/and objects. To resolve this type of query, authors use a *multi-predicate query resolution* algorithm. This algorithm starts by recursively looking for all candidate subjects on each predicate and intersects them at each step before sending back the final results to the query originator. More precisely, let us denote by q the current active sub-query; $\mathcal{R}$ the list of remaining sub-queries that have to be resolved; $\mathcal{C}$ the list of candidate subjects that matches the current active sub-query q and $\mathcal{I}$ the set of intersected subject that matches *all already resolved* sub-queries. Whenever subjects that match the sub-query q are found, they will be added to $\mathcal{C}$. At each forwarding step, an intersection between $\mathcal{C}$ and $\mathcal{I}$ is made in order to keep only subjects that match all already resolved sub-queries. Once *all* results for the current active query are found,the first sub-query in $\mathcal{R}$ is popped to q. Whenever all sub-queries are resolved (i.e., $\mathcal{R} = \{\emptyset\}$) or no results for the current sub-query are reported (i.e., $\mathcal{I} = \{\emptyset\}$), the set $\mathcal{I}$ containing the query results will be sent back to the originator.

While it supports several kinds of queries, RDFPeers has a set of limitations especially during the query resolution phase. This includes the *attribute selectivity* and the restrictions made at the level of supported queries. The attribute selectivity is associated with the choice of the first triple pattern to be resolved. Low selectivity of an attribute leads to a longer computational time to manage the local search as well as a greater bandwidth consumption to fetch results from one node to the other, because many triples will satisfy that constraint. As an example, the predicate 'rdf:type' seems to be less selective, as it can be more frequently used in RDF triples than others. Despite the attribute selectivity parameter having an important impact on the performance of the query resolution algorithm, RDFPeers does not provide a way to estimate such parameter. The pattern selectivity is also considered in [15] where lookups by subject have priority over lookups by predicates and objects and lookups by objects have priority over lookup by predicates. Another issue is related to conjunctive triple pattern queries which are not fully supported and are restricted to conjunctive queries with the same subject so that arbitrary joins are not supported.

Atlas. A similar approach to RDFPeers was presented in the Atlas project [54] which is a P2P system for RDFS data storage and retrieval. It is built on top of Bamboo DHT [76] and uses the RDFPeers query algorithm. Atlas uses RDF Query Language (RQL) [50] as query language and the RDF Update Language (RUL) [28] for RDF metadata insertion, deleting and update.

Subscription processing. Atlas supports a content-based pub/sub paradigm, using algorithms from [59]. In essence, after the submission of an atomic query and its indexing, a peer will wait for triples that satisfy it. Once a new triple is inserted, nodes cooperate together in order to determine which queries are satisfied. Nodes responsible for triples satisfying the subscription create notifications and send them to the subscriber node. Processing of conjunctive subscriptions is a more complicated task since a single triple may partially resolve a query only by satisfying one of its sub-queries. Moreover, as the triples satisfying the query are not necessarily inserted in the network at the same time, nodes need to keep traces of queries that have been already partially satisfied and create notifications only when all sub-queries are satisfied.

Dynamic Semantic Space. In [42], Gu *et al.* propose Dynamic Semantic Space (DSS), a schema-based P2P overlay network where RDF statements are stored based on their semantics. Peers are organized in a ring structure enabling the mapping from a k-dimensional semantic space into a one dimensional semantic space. Peers are grouped into clusters, and clusters having the same semantics are organized into the *same semantic cluster*. While there is no constraint on the overlay topology within the semantic clusters, they themselves form a Chord [83] overlay structure. Each cluster is identified by a `clusterID` given by a k-bit binary string such as $k = x + y$. While the x first bits identify the semantic cluster, denoted SC, the y bits represent the identifier of a cluster c belonging to SC. Therefore, the semantic space infrastructure can have a maximum of 2^x semantic clusters and 2^y clusters per SC. As an example, Figure 3 depicts the architecture of a dynamic semantic space where 4 semantic clusters, SC_1, SC_2, SC_3, SC_4, with 16 clusters in total form the semantic space infrastructure. Peers maintain a set of neighbors in their adjacent clusters as well as in other semantic clusters in order to enable inter-clusters and intra-clusters communication. For instance, as C_2 and C_{16} are the two adjacent clusters of cluster C_1 where the peer p_0 belongs to, p_0 maintains references to peers p_2, p_3 such as $p_3 \in C_2$ ($C_2 \in SC_1$) and $p_2 \in C_{16}$ ($C_{16} \in SC_4$). Moreover, p_0 maintains pointers to other semantic clusters, that is SC_2 and SC_3. DSS provides a load balancing mechanism by controlling the number of peers per cluster and thus the global

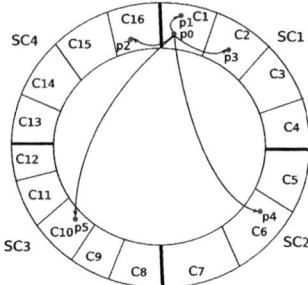

Fig. 3. Example of a Dynamic Semantic Space architecture with 4 semantic clusters and 16 clusters

load. Therefore, a splitting and merging process can occur if the number of nodes per cluster reaches a threshold value.

Data indexing. Authors propose a two-tier ontology-based semantic clustering model: the upper layer defines a set of concepts shared by all peers. A peer needs to define a set of low-layer ontologies and store them locally. Therefore, RDF statements which are semantically similar are grouped together and can thus be retrieved by queries having the same semantics. To join the network, a peer has to merge the upper layer ontology with its local ontologies and create RDF data instances which will be added to the merged ontology to form its local knowledge base. A mapping of its RDF data to one or more semantic clusters is performed and the peer joins the most semantically related cluster.

Query processing. The query processing mechanism is performed in two steps. First, when the query, expressed in RDF-Data Query Language (RDQL), is received, the peer pre-processes the query to get the information about the semantic cluster SC. Second, when the query reaches the targeted SC, it will be flooded to all peers within SC.

Subscription processing. DSS also supports a content-based pub/sub mechanism. Once a subscription request is generated it will be mapped to the corresponding SC. From there, it will be forwarded to every node within SC. Upon receipt of the subscription request, a peer checks for a local matching of the subscription against its local RDF data. After a successful matching, it will send back the results to the subscriber(s). If a modification occurs on the local data, the producer peer notifies subscriber node(s) by sending notifications that follow exactly the reverse path of the corresponding subscription. Whenever a peer wants to unsubscribe to a query, an unsubscription request will be sent directly to the relevant producers.

PAGE. Put And Get Everywhere (PAGE) presented in [33] is a RDF distributed repository based on Bamboo DHT [76]. The RDF data model introduced in this work extends the standard RDF data model by introducing the notion of `context`. This concept is application-dependent. For instance, in the information integration use case, the context is the URI of the file or the repository from which a RDF triple originated. In other scenarios, RDF triples sharing the same semantics can have the same context. Therefore, in PAGE, each RDF triple, denoted by t=(s,p,o), is associated with a context, labeled c. The RDF triple combined with the context forms what is called a quad. Each quad is indexed six times and identified by an ID that is built by concatenating the hash values of its elements (s,p,o,c).

Data indexing. PAGE associates to each index an index code. This field identifies the used index (i.e., {spoc,cp,ocs,poc,csp,os}) for building the quad ID. For instance, suppose that one wants to store the quad quad=(x,y,z,w). Assume that the hashed values of its elements are given by hash(x)=1, hash(y)= F, hash(z)=A, hash(w)=7; and indexes codes by: idx(spoc)=1, idx(cp)=4, idx(ocs)=6, idx (poc)= 9, idx(csp)= C, idx(os)=E. The concatenation of '1FA7' with idx (spoc)=1 results to ID='11FA7'. The quad will then be stored on the node that has the numerically closest identifier to ID. The same operation is performed for each of these six indexes giving at each time the node identifier where the quad will be stored.

Query processing. The query processing algorithm starts by associating to each access pattern specified in the query an identifier ID as explained earlier. A *mask* property is used to specify the number of elements fixed in the query. Moreover, the query is controlled by the number of *hops* which denotes the number of digits from the query ID that have been already considered for the query evaluation during the routing process. Take back the example discussed above and consider the access pattern q=(s,p,?,?) which looks, for the given subject s and predicate p, for all object values for any context. This access pattern requires the index (idx(spoc)=1) to be resolved. For unknown parts a sequence of 0 is assigned. The query will then be converted to 11F00/3. The mask value here is 3 meaning that the first three digits of the identifier are fixed. Thereby, only the *object* and the *context* will be considered during the query processing.

3.2   Cube-Based Approaches

Edutella.   Nejdl *et. al* propose a RDF-based P2P architecture called Edutella [72] which makes use of super-peers. In this kind of topology, a set of nodes are selected to form the super-peer network, building up the "backbone" of the P2P network, while the other peers connect in a star-like topology to those super-peers. Super-peers form a so-called *HyperCuP* topology [80] and are responsible for the query processing. In this organization, each super-peer can be seen as the root of a spanning tree which is used for query routing, broadcasting and indices updating. Each peer sends indices of its data to its super-peer. Edutella implements a schema-based P2P system where the system is aware of schema information and takes it into consideration for query optimization and routing. Data indices can be stored in different granularities, like schema, property, property value or property value range. However, this index contains peer indices instead of data entries. This kind of index is called a Super-Peer-Peer

(SP-P) index. For example, a query with a given schema received by a super-peer will be only forwarded to peers supporting that schema. This allows to narrow down the number of peers to which the query will be broadcasted. In addition to the SP-P index, the super-peers maintain Super-Peer-Super-Peer (SP-SP) indices in order to share their index information with other super-peers. Accordingly, queries are forwarded to super-peer neighbors based on those indices. Edutella supports the RDF Query Exchange Language (RDF-QEL).

Data indexing. When a new peer registers to a super-peer, labeled sp, it advertises the necessary schema information to sp. Sp checks this new schema against its SP-P index entries. If there is new information that has to be added in the SP-P index, sp broadcasts the new peer advertisement to its super-peers neighbors in order to update their SP-SP indices. In a similar way, indexes have to be updated whenever a peer leaves the network.

Query processing. Query routing in Edutella is improved by using *similarity-based clustering strategy* at the super-peer level to avoid as much as possible the broadcasting during the query routing phase. This approach tries to integrate new peers with existing peers that have similar characteristics based, for example, on a topic ontology shared by them.

RDFCube. Monato *et al.* propose RDFCube [67], an indexing scheme of RDF-Peers (Section 3.1) based on a three-dimensional CAN coordinate space. This space is made of a set of cubes, with the same size, called *cells*. Each cell contains an *existence-flag*, labeled e-flag, indicating the presence (e-flag=1) or the absence (e-flag=0) of a triple within the cell. The set of consecutive cells belonging to a line parallel to a given axis forms a *cell sequence*. Cells belonging to the same plane perpendicular to an axis form the *cell matrix*.

Data indexing. Once a RDF triple t=(s,p,o) is received, it will be mapped to the cell where the point p with the coordinates (hash(s),hash(p),hash(o)) belongs to.

Query processing. As for RDF triples, the query is also mapped into a cell or a plane of RDFCube based on the hash values of its constant part(s). Consequently, the set of cells including the line or the plane where the query is mapped are the candidate cells containing the desired answers. Note that RDFCube does not store any RDF triples but it stores bit information in the form of e-flags. The interaction between RDFCube and RDFPeers is as follows: RDFCube is used to store (cell matrixID, bit matrix) pairs such as the matrixID is a matrix identifier and represents the key in the DHT terminology, while bit matrix is its associated value. RDFPeers stores the triples associated with the bit information. This bit information is basically used to speed up the processing of a *join query* by performing an AND operation between bits and transferring only the relevant triples. As a result, this scheme reduces the amount of data that has to be transferred among nodes.

3.3   Tree-Based Approaches

GridVine. GridVine [31] is a distributed data management infrastructure based on P-Grid DHT [8] at the overlay layer and which maintains a *Semantic Mediation Layer* on top of it as it is shown in Figure 4. GridVine enables distributed search, persistent storage and semantic integration of RDF data. The upper layer takes advantage of the overlay architecture to efficiently manage heterogeneous data including schemas and schema mappings.

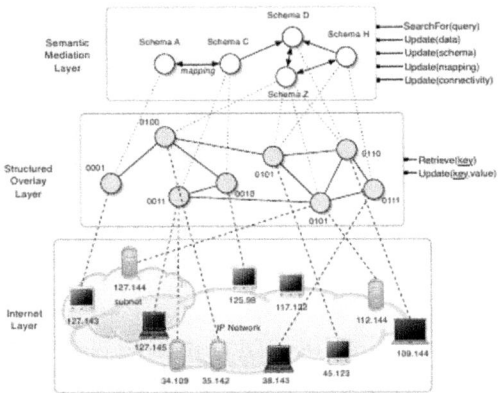

Fig. 4. GridVine architecture incarnated in the semantic mediation layer and the structured overlay layer (taken from [31])

Data indexing. As in RDFPeers [23], GridVine indexes triples three times at the P-Grid layer based on their subjects, objects and predicates. As an example, for a given triple t, the insert(t) operation results into insert(hash(s), t), insert(hash(p), t) and insert(hash(o), t). In that way, a triple inserted at the mediation layer triggers three insert operations at the overlay layer. GridVine also supports the sharing of schemas. Each schema is associated with a unique key which is the concatenation of the logical address of the peer posting the schema, say $\pi(p)$, with the schema name, Schema_Name. Therefore, a schema indexing operation may take the form of Update($\pi(p)$: hash(Schema_Name), Schema _Defintion). In order to integrate all semantically related yet syntactically heterogeneous data shared by peers at the P-Grid level, GridVine supports the definition of pairwise semantic mappings. The mapping allows the reformulation of a query against a given schema into a new query posed against a semantically similar schema. The schema mapping uses Web Ontology Language (OWL) [12] statements to relate two similar schemas.

Query processing. Lookup operations are performed by hashing the constant part(s) of the triple pattern. Once the key space is reached, the query will be forwarded to peers responsible for that key space. As in P-Grid, the search complexity of atomic triples pattern is $\mathcal{O}(\log(\Pi))$ such as Π is the entire key partition.

Note that atomic queries where none constant part is fixed are not supported by the query processing scheme. GridVine is also able to handle disjunctive and conjunctive queries by iteratively resolving each triple pattern in the query and perform distributed joins across the network. Therefore, the system query processing capabilities are very similar to RDFPeers, but GridVine takes also into consideration the schema heterogeneity which is not addressed by RDFPeers.

UniStore. Karnstedt *et al.* present their vision of a universal data storage at the Internet-scale for triples' repository through the UniStore project [49]. Their solution is based on P-Grid [8], on top of which they built a triple storage layer that enables the storage of data as well as metadata.

Data indexing. Authors chose to adopt a universal relation model allowing schema-independent query formulation. In order to speed up and take advantage of the underlying features of the DHT for fast lookups, all attributes are indexed. When dealing with relational data, each tuple $(\texttt{OID}, v_1, \ldots, v_x)$ of a given relation schema $\texttt{R}(\texttt{A}_1, \ldots, \texttt{A}_x)$ is stored in the form of x triples: $(\texttt{OID}, \texttt{A}_1, v_1), \ldots, (\texttt{OID}, \texttt{A}_x, v_x)$ where $\texttt{OID}$ is a unique key, $\texttt{A}_i$ an attribute name and v_i its corresponding value. The $\texttt{OID}$ field is used to group a set of triples for a logical tuple. Each triple is indexed on the $\texttt{OID}$, the concatenation of $\texttt{A}_i$ and v_i ($\texttt{A}_i\#v_i$) and finally on the value v_i.

Query processing. UniStore allows the use of a query language called Vertical Query Language (VQL) which directly derives from SPARQL. This language supports DB-like queries as well as IR-like (Information Retrieval) queries. It comes with an algebra supporting traditional "relational" operators as well as operators needed to query the distributed triple storage. It offers basic constructs such as SELECT, ORDER BY, etc. as well as advanced ones such as SKYLINE OF. Furthermore, in order to support large-scale and heterogeneous data collection, the language was extended with similarity operators (e.g., similarity join) and ranking operators (e.g., skyline queries). Operators can be applied to all levels of data (e.g., instance, schema and metadata). Each of these logical operators have a "physical" implementation which only rely on functionality provided by the overlay system (key lookup, multicasts, etc.) and they differ in the kind of used indexes, applied routing strategy, parallelism, etc. These physical operators are used to build complex query plans, which in turn are used to determine worst-case guarantees (almost all of them being logarithmic) as well as predict exact costs. This allows the system to derive a cost model for choosing concrete query plans, which is repeatedly applied at each peer involved in the query resolution and thus resulting in an adaptive query processing approach.

3.4   Generic DHT-Based Approaches

Battré et al. In [16], authors propose a data storage and retrieval strategy for DHT-based RDF store. They use similar indexing mechanisms as in [9, 23, 67], the proposed approach indexes the RDF triple by hashing its subject, predicate and object. It takes into consideration RDFS reasoning on top of DHTs by applying RDFS reasoning rules.

Data indexing. The main characteristic of this approach compared to others is that nodes host different RDF repositories, making a distinction between local and incoming knowledge. Each RDF repository is used for a given purpose:

- **Local triples repository** stores triples that originate from each node. A local triple is disseminated in the network by calculating the hash values of its subject, predicate and object and sending it to nodes responsible for the appropriate parts of the DHT key space.
- **Received triples repository** stores the incoming triples sent by nodes.
- **Replica repository** ensures *triple availability* under high peer churn. The node with the numerically closest ID to the hash value of a triple becomes the root node of the replica set. This node is responsible for sending all triples in its received repository to the replica nodes.
- **Generated triples repository** stores triples that are originated from applying forward chaining rules on the received triples repository, and they are then disseminated as local triples to the target nodes. This repository is used for RDFS reasoning.

In order to keep the content of the received triples repository up-to-date, specially under node leaving or crashing, triples are associated with an expiration date. Therefore, the peer responsible of that triple is in charge of continuously sending update messages. If the triple expiration time is not refreshed by the triple owner, it will be eventually removed from these repositories. This approach takes care of load balancing issues in particular for uneven key distribution. For instance, the DHT may store many triples with the same predicate 'rdf:type'. As the subject, predicate and object will be hashed, the node responsible for the hash(rdf:type) is a target of a high load. Such situation is managed by building an overlay tree over the DHT in order to balance the overloaded nodes.

Query processing. In another work [45], one of the authors proposes a RDF query algorithm with optimizations based on a look-ahead technique and on Bloom filters [21]. Knowledge and queries are respectively represented as *model* and *query* directed graphs. The query processing algorithm basically performs a matching between the query graph and the model graph. On one side, there is the candidate set which contains all existing triples, and on the other side, there is a candidate set containing the variables. These two sets are mutually dependent, therefore a refinement procedure has to be performed to retrieve results for a query. This refinement proceeds in two steps. First, starting from the variable's candidate set, a comparison is done with the candidate sets for each triple where the variable occurs. If a candidate does not occur within the triple candidate set, it has to be removed from the variable candidate set. The second step goes the other way around, that is, it looks at the candidate set for all the triples and removes every candidate where there is no matching value within the variable's candidate set.

The look-ahead optimization aims at finding better paths through the query graph by taking into account result set sizes per lookup instead of the number of lookups. This yields fewer candidates to transfer but the trade-off is that it incurs more lookups.The second optimization, using Bloom filters, works as follows: consider candidates for a triple (a, v_2, v_3), where a is a fixed value and v_2 and v_3 are variables. When retrieving the candidates by looking up using the fixed value a, it may happen that the querying node might already have candidates for the two variables. The queried node can reduce the results' sets with the knowledge of sets v_2 and v_3. However, those sets may be large, for this reason the authors use Bloom filters to reduce the set representation. The counterpart of using Bloom filters, is that they yield false positives, i.e., the final results sets which will be transferred may contain non-matching results. A final refinement iteration will be done locally which will remove those candidates and thus ensuring the correctness of the query results.

Query Chain and Spread by Value Algorithms. In [59], Liarou *et al.* propose the Query Chain (QC) and Spread by Value (SBV) algorithms to evaluate conjunctive queries over structured overlays.

QC - Data indexing. As in RDFPeers [23], in QC algorithm, a RDF triple is indexed to three nodes. More precisely, for a node that wants to publish a triple t such as t=(s,p,o), the index identifiers of t is computed by applying a hash function on s, p and o. Identifiers hash(s), hash(p) and hash(o) are used to locate the nodes that will then store the triple t.

QC - Query processing. In this algorithm, the query is evaluated by a *chain* of nodes. Intermediate results flow through the nodes of this chain and the last node in the chain delivers the result back to the query's originator. More precisely, the query initiator, denoted by n, issues a query q composed of q_1, $q_2, \ldots, q_i$ patterns and forms a *query chain* by sending each triple pattern to a (possibly) different node, based on the hash value of the constant part(s) of each pattern. For each of the identified nodes, the message QEval(q, i, R, IP(x)) will be sent such that q is the query to be evaluated, i the index of the pattern that will managed by the target node, IP the address of the query's originator x and R an intermediate relation to hold intermediate results. When there is more than one constant part in the triple pattern, subject will be chosen over object over predicate in order to determine the node responsible for resolving this triple. While the query evaluation order can greatly affect the algorithm performance including the network traffic and the query processing load, the authors adopt the default order for which the triple patterns appear in the query.

SBV - Data indexing. In the SBV algorithm, each tripl $t = (s, p, o)$ is stored at the successor nodes of the identifiers hash(s), hash(p), hash(o), hash(s+p), hash(s+o), hash(p+o) and hash(s+p+o) where the '+' operator denotes the concatenation of string values. Using triple replication, the algorithm aims to achieve a better query load distribution at the expense of more storage space.

SBV - Query processing. SBV extends the QC algorithm in the sense that the query is processed by multiple chains of nodes. Nodes at the leaf level of these chains will send back results to the originator. More precisely, a node posing a conjunctive query q in the form of $q_1 \wedge \ldots \wedge q_k$ sends q to a node n_1 that is able to evaluate the first triple pattern q_1. From this point on, the query plan produced by SBV is created dynamically by exploiting the values of the matching triples that nodes find at each step. As an example, a node n_1 will use the values found locally that matches q_1, to bind the variables of $q_2 \wedge \ldots \wedge q_k$ that are in common with q_1 and produce a new set of queries that will jointly determine the answer to the query's originator. Unlike the query chain algorithm, to achieve a better distribution of the query processing load, if there are multiple constants in the triple pattern, the concatenation of all constant parts is used to identify nodes that will process the query.

The performance of both QC and SBV algorithms can be improved through the caching of the IP addresses of contributing peers. This information can be used to route similar queries and thus reduce the network traffic and query response time. A similar algorithm is presented in [14] where conjunctive queries are resolved iteratively: starting by a single triple pattern and performing a lookup operation to find possible results of the current active triple pattern. These results will be extended with the resolution of the second triple pattern and an intersection of the current results with previous ones will be done. The procedure is repeated until all triple patterns are resolved. Unlike Query Chain algorithm discussed in [59], where intermediate results for a conjunctive query resolution are usually sent through the network to nodes that store the appropriate data, Battré in [14], combines both the fetching and the intermediate results forwarding approaches. Such decision is taken at runtime and is based on the estimated traffic for each data integration techniques.

3.5 P2P-Based Publish/Subscribe Systems for RDF Data Storage and Retrieval

Cai et al. Cai *et al.* have proposed in [24] a content-based pub/sub layer extension atop their RDFPeers infrastructure [23] (Section 3.1). In their system, each peer n maintains a list of local subscriptions, along with the following information: triple patterns, the subscriber node identifier, a requested notification frequency and the requested subscription expiration date. Once the node n stores (upon insertion) or removes a triple locally, it evaluates the matching subscription queries. This is also done after a triple update or when a collection of several matches for a given subscription is made. Afterwards, p notifies the subscriber by sending matched triples.

Subscription processing. The subscription mechanisms support *atomic, disjunctive, range* and *conjunctive multi-predicate* queries[1]. The basic subscription mechanism for *disjunctive* and *range* queries is similar to the one used for atomic queries, except that a subscription request is stored by all nodes that fall into the hashed identifiers of the lower and the upper range bounds. The subscription scheme to manage *conjunctive multi-predicate* subscriptions is similar to the query chain algorithm discussed in Section 3.4. Initially, the subscription request S_r will be routed to the node n corresponding to the first triple pattern s in the

---

[1] Not implemented at the time of the writing of their paper and some combinations of conjunctive and disjunctive queries are not supported.

subscription. Once the node n processes the first conjunct s, it removes s from S_r and stores it locally. In addition, the node n also stores the hash value of the next pattern in S_r. The node, receiving the subscription request (i.e., $S_r \setminus \{s\}$) sent by p stores the next pattern and routes the remaining patterns towards the appropriate peer and so on and so forth. The node that stores the last pattern in the subscription S_r will also store the subscription request's originator. Therefore, whenever new triples are matched by the first pattern, they will be forwarded to the second node in the chain. The second and subsequent nodes will only further forward those triples if they also match their local filtering criterion. Moreover, the authors propose an extension to support highly skewed subscriptions patterns, to avoid having a majority of nodes with few if any subscribers while a handful of node attracts a lot of subscribers. This extension promotes a proportional notification's dissemination scheme to the number of subscribers. The authors provide two ways to do so: (i) similar subscriptions from different subscribers are aggregated and reduced into a single entry with multiple subscribers and (ii) when a certain threshold of subscribers is exceeded, a node will maintain a "repeater nodes" list which is used to distribute the load of the notification propagation among nodes. In order to cope with churn and failure, a replication mechanism, directly inherited from MAAN [22], along with a repair protocol for the subscriptions and the data is provided but only for atomic queries. Replication can be tuned using a `Replica_Factor` parameter, thus, the subscription list will be replicated to the next `Replica_Factor` nodes in the identifier space. The same principle can be applied to RDF triples. To manage situations where a user leaves while his subscriptions are still active, each subscription is characterized by a *maximum duration parameter* which controls its lifetime.

Chirita et al. In [10], Chirita *et al.* propose a set of algorithms to manage publications, subscriptions and notifications on top of super-peer like topology as in Edutella [72] (Section 3.2). The authors formalized the basic concepts of their content-based pub/sub system using a *typed first-order* language. Such formalization enables a precise reasoning on how subscriptions, advertisements and publications are managed in such systems. The formalism is further used by the authors to introduce the *subscription subsumption* optimization technique, which we will explain hereafter.

Advertisement indexing and processing. In this model, once a peer n connects to its super-peer node, denoted sp, it compulsorily advertises the resources it can offer in the future by sending advertisements to sp. These advertisements are useful to super-peers in order to build *advertisement routing indices* that will further be used for subscriptions processing. Advertisements contain one or more elements which can either be a schema identifier (e.g.,<dc>,<lom>), a property/value pair (e.g.,{<dc:year>,2010}) or a single property (e.g.,<dc:year>). Consequently, three levels of indexing are managed:

- **Schema level.** This level of indexing contains information about RDF schemas supported by the peers. Each schema is uniquely identified by a

URI. The schema indexing can be used, for example, to constrict the subscriptions forwarding only to the peers supporting that schema.

- **Property/Value level.** This second level indexes peers providing the resources rather than the property itself. It is mainly used to reduce the network traffic.
- **Property level.** This level of indexing is useful when a peer chooses to use specific properties/attributes from one or a set of schemas. Thus, the property index contains properties, identified by `namespace/schemaID` in addition to the property name. Each property points to one or more peers that support them.

Advertisements are selectively broadcasted from a super-peer to its super-peer neighbors. The advertisement update process is triggered if a peer joins or leaves the network or whenever the data that it is responsible for is modified. Therefore, references associated to peers and indexed at the super-peer level also have to be updated.

Subscription processing. Each super-peer manages a subscription in a partially ordered set (poset). Subscriptions sent by a peer n are inserted into the poset of its super-peer sp. As in the SIENA system [25], the poset consists of a set of subscriptions. The subscriptions' poset is actually a hierarchical structure of subscriptions which captures the concept of subscription subsumption, that is, when the result of a subscription q_1 is a subset of the result subscription q_2, we say that q_2 *subsumes* q_1. When new subscriptions are subsumed by previously forwarded ones, the super-peer does not forward these new subscriptions to its super-peer neighbors. However, consider an advertisement Adv_1 sent by sp_2 to sp_1. If sp_1 has already received a subscription S (by another peer) and there is a match with Adv_1, then sp_1 informs sp_2 not to forward any advertisement of this kind anymore, because the subscription responsible for S subsumes Adv_1.

When a new notification *not* arrives at a super-peer sp, sp checks for the subscriptions satisfying *not* in its local subscriptions poset in order to forward them to the subscriber nodes. This approach handles the notifications under peer churn. When a peer n associated to a super-peer sp leaves the network, sp puts the subscriptions of n in a buffer for a period of time in order to receive notifications sent to n. Once n reconnects to sp, these notifications will then be forwarded to it. The authors also make use of cryptographic algorithms to provide unique identifiers to peers. After a leave (or a crash), a node may have a different IP address. This unique identifier will be used to retrieve waiting subscription results.

Single and Multiple Query Chain Algorithms. Liarou *et al.* have proposed in [58] a content-based pub/sub system, which mainly focuses on conjunctive multi-predicate queries. They provide two *DHT-agnostic* algorithms, namely, the *Single Query Chain* (SQC) and the *Multiple Query Chains* (MQC), that extends the Query Chain algorithm detailed in Section 3.4.

Subscription processing. In their pub/sub system, each triple t is character-ized by a *published time* parameter, labeled $\texttt{pubT}(t)$. Moreover, each subscription S such as $S = s_1 \wedge s_2 \wedge \ldots \wedge s_n$ is identified by a $\texttt{key}(S)$ and has a timestamp indi-cating the *subscription time*, denoted by $\texttt{subscrT}(S)$. As a result, s_i is also char-acterized by a subscription time such as $\texttt{subscrT}(s_i) = \texttt{subscrT}(S)$. Therefore, a triple t can satisfy a subscription s of S *iff* $\texttt{subscrT}(s) < \texttt{pubT}(t)$. Subscrip-tions and notifications forwarding policies are similar to algorithms proposed in [59]. In the case of SQC, for each query, a single query chain is created at node r upon receipt of query S whereas MQC goes a step further. In MQC, first, a query S is indexed to a single node r according to one of S's sub-queries. Then, each time a triple arriving at r satisfies this sub-query, the subject is used to rewrite S and thus, for each different rewritten query, a query chain is created, yielding multiple query chains. This has the benefit of achieving a better load distribution than SQC. For optimization purposes, the authors propose a query clustering mechanism where similar subscriptions are grouped together. For in-stance, triples which have been indexed on node n using the same predicate p will be answered when a new triple with the predicate p is inserted. In such a scenario, only one matching operation has to be performed whenever such a triple is inserted.

Continuous Query Chain and Spread-By-Value Algorithms. In another work [57], Liarou *et al.* have proposed an extension of their two algorithms pre-sented in [59] (QC and SBV). By introducing a *continuous* version of these algorithms, thus becoming CQC (*Continuous Query Chain*) and CSBV (*Con-tinuous Spread By Value*).

CQC - Subscription processing. A node n that wants to subscribe to a con-junctive query $q = [q_1, \ldots, q_k]$ will do so by indexing each triple pattern q_j to a different node n_j. Thus, each of these nodes will be responsible for processing their part of the query q_j and will be a part of a *query chain* of q. Each node indexes the triple patterns as in QC, explained in Section 3.4. When a node receives a newly indexed triple t, it will check any relevant indexed queries in its *query table (QT)*. If any match is found, there is a *valuation*[2] v over the variables of q_j such that $v(q_j) = t$. (i). If it is the first node in the chain, it forwards the valuation v (holding a partial answer to q) to the next node in the chain, n_i. When receiving this *intermediate result*, n_i will apply the received valuation and compute a new valuation w to the pattern q_i it holds, resulting in $q'_i = w(q_i)$. Then, it will try to find triples matching q'_i in its *triple table (TT)* that have already arrived. If there is a match, a new intermediate result is produced, a new valuation $w' = w \cup v$. This new valuation is then passed along the query chain and stored in an *intermediate result table (IRT)*, which is used when new triples arrive. When the last node in the chain receives a set of intermediate results, it will check its *TT* and if matches are found, an *answer* to the query q is sent back to the node that originally posed q. (ii). If the node is not the first in the query chain, it will store the new triple in its *TT* and search in its *IRT* to see if

---

[2] Concept used to denote values satisfying a query.

an evaluation of the query that has been suspended can now continue due to t that has just arrived. For each intermediate result w found, a new valuation w' is computed and forwarded to the next node in the query chain.

CSBV -Subscription processing. This algorithm actually extends CQC in a way that it achieves a better distribution of the query processing load. In CQC, a query chain with a fixed number of participants is created upon submission of a query q, whereas in CSBV no query chain is created. Instead q is indexed *only to one node* which will be responsible for one of the triple patterns of q. A node n can use the valuation v to rewrite $q = [q_1, \ldots, q_k]$ with fewer conjuncts $q' = [v(q_2), \ldots, v(q_n)]$ and decides *on the fly* the next node that will undertake the query processing. Because q' is conjunctive like q, its processing proceeds in a similar manner. Depending on the triples that trigger q_i, a node can have multiple next nodes for the query processing. Thus, the responsibility of evaluating the next triple pattern of q is distributed to multiple nodes compared to just one node in CQC, leading to a better load distribution. Query indexing follows the same heuristics used in CQC, with the difference that if the query has multiple constants, a combination of all the constant parts will be used to index the query, following the triple indexing of SBV presented in Section 3.4.

When receiving a new triple t, this basically proceeds as in CQC (in terms of checking if a match is found for the pattern) with the only exception that the forwarding of the intermediate results is dynamic because computed on the fly. Intermediate results are processed in the same logic as in CQC with the difference that instead of forwarding a single intermediate result to the next node, a *set* of intermediate results is created and delivered possibly to different nodes. These two algorithms come with communication primitives that enable messages to be sent in bulk. Naturally, each step of these algorithms comes at a certain cost, but some improvement techniques are reused from QC and SBV such as IP caching used to maintain the address of the next node in the chain the peer should send the intermediate results to, or also a query chain optimization scheme which tries to find an optimized nodes' order based on specific metrics such as the rate of published triples that trigger a triple pattern.

Ranger et al. In [74], the authors introduce an information sharing platform based on Scribe [26], a topic-based pub/sub system with which they use queries as topics. Interestingly, the algorithm they propose does not index data *a priori.* Instead, their scheme relies upon finding results from scratch with redundant caching and cached lookups mechanisms. This is done by taking advantage of the Scribe infrastructure, that is, the possibility to subscribe to a topic and to lookup within the group of the topic. Frequent queries are cached effectively as they occur, remaining active as long as clients are reading from it. Queries, which can be either atomic or complex, are expressed in a SPARQL dialect.

Subscription processing. Queries are translated into trees of either atomic or complex sub-queries. An atomic query is a simple, independent query that all peers can execute on their local content since it does not depend on the results from other peers. A complex query resorts to results from other queries to

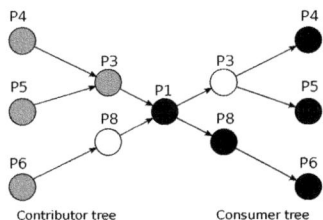

Fig. 5. Contributor and consumer pipelines resulting from P1 performing a query. Gray nodes are contributors to query Q; white nodes are forwarders, that is, nodes which help in the dissemination even if they are not interested in the content; black nodes are consumers of the results integrated by P1 (figure taken and redrawn from [74]).

produce its results, either by aggregation, filtering or calculation. Since all peers can execute arbitrary complex queries, the exact distinction between atomic and complex queries is somehow fuzzy. It is usually unfeasible to determine if the query should be run locally. This depends on whether objects are stored by a single peer, or if parts of objects are scattered across different peers. This also depends on the application built on top of this scheme. In order to circumvent this ambiguity, the algorithm co-locates predicates from designated namespaces (e.g., dc,rdf,type,etc.). This means that for a same subject, all predicates from a co-located namespace will be available on the same peer. When a peer performs a query, it first looks up in the group for a peer p that already knows (or is interested to know) the query's result. If p is found, the peer joins the so called *consumer* pipeline tree rooted at a producer of the query results. When a peer matches an atomic query, it establishes a *contributor* pipeline tree rooted at itself. When consumers ask for results, the producing peer will read/combine results from the contributor pipeline and forward them to the consumers. Figure 5 shows the result pipelines established when a peer P1 performs a query. Within a group, whenever a peer is processing an atomic query, it will broadcast the fact to other peers.When a peer receives a broadcast message, if it has relevant content or if the query is *live*, that is if the originator wants to be able to wait for new results as they happen, it will be forced to contribute. The sole reason for this is to avoid ignoring any potential source of results. The algorithm makes use of Pastry's built-in mechanisms [78] to become aware of intentional or accidental departures, and the contributing or forwarding peer will be asked to re-send the message and thus repairing the tree with minimal losses.

4   Summary and Discussions

The combination of P2P technologies with RDF data has become a very active research area aiming at sharing and processing large amounts of data at the scale of the Internet. In addition to the inherent challenges related to such large scale infrastructure (e.g., network partition, topology maintenance, fault-tolerance, etc.), enabling complex querying of RDF data on top of such infrastructures requires advanced mechanisms, especially from a data indexing and query processing point of views. The presented works are summarized in Table 2 and hereafter we will discuss some challenges surrounding this combination and point out some improvements with respect to the underlying overlays, the way the RDF data is indexed and finally how queries are processed.

Table 1. Comparison of the different presented approaches in terms of indexing mechanism, the routing scheme and the data replication property

	Topology	Hashing	Routing scheme	Replication
Ring	RDFPeers [23]	✓	Index-based	✓
	Atlas [54]	✓	Index-based	✓
	DSS [42]	✗	Semantic mapping	✗
	PAGE [33]	✓	Index-based	✓
TreeCube	Edutella [71]	✗	Semantic mapping	✗
	RDFCube [67]	✓	Index-based	
	GridVine [9]	✓	Semantic mapping and Index-based	✓
	Unistore [49]	✓	Index-based	✓
Generic DHT	Battré et al. [16, 45]	✓	Index-based	✓
	QC & SBV [59]	✓	Index-based	✓

Overlay topology. As we have already seen, a lot of work has been carried out towards a scalable distributed infrastructure of RDF data storage and retrieval. Most of the presented approaches choose the structured P2P networks as infrastructure basis (e.g., [9, 16, 23, 33, 42, 44, 54, 59, 67, 72]). Such choice was motivated by the fact that, first, as it is already argued in [62], structured P2P overlays provide a practical substrate for building scalable distributed applications due to the guarantees that they provide. Secondly, the structural nature of the RDF data model influences the choice of the topology as it can be noticed through the data indexing model of most of the presented approaches. We detailed how each work indexes RDF data on top of their overlay and associated topology (that is why we chose a topology oriented classification in Section 3). However there is no clear evidence from the state of the art study that a specific overlay topology is the most suitable for handling RDF data.

Data indexing. As we have already seen, a lot of work has been carried out towards a scalable distributed infrastructure of RDF data storage and retrieval. Most of the presented approaches choose the structured P2P networks as infrastructure basis (e.g., [9, 16, 23, 33, 42, 44, 54, 59, 67, 72]). Such choice was motivated by the fact that, first, as it is already argued in [62], structured P2P overlays provide a practical substrate for building scalable distributed applications due to the guarantees that they provide.

Secondly, the structural nature of the RDF data model influences the choice of the topology as it can be noticed through the data indexing model of most of the presented approaches.

Table 2. Qualitative comparison of P2P systems for RDF data storage and retrieval: s, p, o respectively refer to the subject, the predicate and the object of a RDF triple and c denotes the *context* of the application.

| Topology | Contribution details | | | |
	Data indexing	Indexing particularities	Query language and Supported queries	Query processing
Ring-based — RDFPeers [23] MAAN [22]	Three-times RDF triple indexing.	Locality preserving hash function: hash(s), hash(p), hash(o).	Atomic, conjunctive, disjunctive queries with the *same* subject	Recursive multi-predicate query resolution.
Ring-based — Atlas [54] Bamboo[76]	RDFPeers indexing scheme.	Hashing.	RQL; conjuctive triple pattern queries.	RDFPeers query processing algorithm; Subscriptions and matching mechanisms derive from QC and SBV [59].
Ring-based — DSS [42] Chord [83]	Semantic clustering.	Semantic-based RDF data organization.	RDQL.	Routing within the most suitable semantic cluster; flooding-based approach inside the cluster; Subscriptions mapping to a semantic cluster and forwarded to all nodes within the cluster, RDF mapping to the most suitable semantic cluster.
Ring-based — PAGE [33] Bamboo[76]	Six-times quad indexing.	quads(s,p,o,c) indexing instead of RDF triples; Associate an ID to each quad.	YARS Query Language.	Use mask and hops properties during the query resolution.
Cube-based — Edutella [71] HyperCuP [80]	Semantic clustering at the Super-Peer level.	Super-Peer-Super-Peer (SP-SP) and Super-Peer-Peer (SP-P) indices.	RDF-QEL.	Lookup at the super-peer level based on SP-SP indices; Lookup inside the cluster based on SP-P indices.
Cube-based — RDFCube [67] CAN [75]	RDFPeers indexing scheme.	Store bit information about the existence of RDF triples.	Conjunctive queries.	Query mapping into a cell of RDFCube through hashing; Speed up *join queries* by performing an AND operation on bit information.
Tree-based — GridVine [9] P-Grid [8]	Three-times RDF triple indexing at P-Grid layer; Schema indexing at the semantic mediation layer.	Order-preserving hashing; Semantic mediation layer overlay atop P-Grid.	OWL; atomic, conjunctive, disjunctive queries.	Schema mapping; Iterative query resolution.
Tree-based — Unistore [49] P-Grid [8]	Three-times indexing on the OID, the concatenation of A_1 and v_1, the value v_1.	a tuple (OID, $v_1, \ldots, v_n$) of a given relation schema $R(A_1, \ldots, A_n)$ is stored as n triples:(OID, A_1, v_1),...,(OID, A_n, v_n).	VQL; database and information retrieval-like queries (similarity, ranking, etc.).	Schema-independent query formulation, Semantic layer atop P-Grid [8]; Adaptive cost-aware query processing approach.
Generic-DHT-based — Battré et al. [16, 45]	RDFPeers-like indexing mechanism.	Different RDF repositories hosted by each node local/received/replica/generated repositories.	Not specified.	Knowledge and query graph-based models; Two-phase refinement procedure for query resolution; A look-ahead and Bloom filters enhancement techniques.
Generic-DHT-based — QC and SBV [59]	Three-times (QC) and seven-times (SBV) RDF triple indexing.	QC: hash(s), hash(p), hash(o); SVB: hash(s),hash(p), hash(o), hash(s+p), hash(s+o), hash(p+o), hash(s+p+o).	The query language is not clearly specified; atomic and conjunctive queries.	Query Chain(QC): Query evaluation by a *chain* of nodes and intermediate results flow through the nodes of the chain; last node in the chain delivers the result back to the query's originator; Spread by Value (SBV): construct multiple chains for the same query.

Several approaches [16, 23, 44, 54, 67], while based on different overlay topologies, share almost the same data indexing model by *hashing* the RDF triple elements (see Table 1) The main advantage of such indexing strategy is that triples with the same subject, object and predicate are stored on the same node and thus can be searched locally without needing to be collected from different data sources. However, nodes responsible for overly popular triples (e.g., 'rdf:type', 'dc:title'), can be easily overloaded resulting in poor performances. One possible solution for this issue would be to use multiple hash functions to ensure a better load distribution as proposed in [75] and recently in [69].

The second category of RDF-based P2P approaches harnesses the semantic of RDF data either to build a "semantic" layer on top of the P2P overlay (e.g., GridVine [31]) or to adopt a semantic clustering approach and organize the P2P layer as function of the semantic of the stored data (e.g., Edutella [42, 72]). Others have extended the basic RDF data model by adding the "context" concept as in Page [33]. By taking into consideration data semantics in the data indexing phase, these approaches try to improve the data lookup. However, this needs an additional effort to maintain the mapping between the semantic and the overlay levels. Most of the presented works aim at adding data *availability* feature to the RDF storage infrastructure through *data replication* (e.g, [16, 23, 31, 54, 59]). However, data replication further raises several issues. Thereof, three main challenges have to be taken into consideration: which data items have to be replicated; where to place replicas and finally how to keep them consistent. In P2P systems there has been a lot of work on managing data replication. In [55], Ktari *et al.* investigated the performance of several replication strategies under DHTs systems including neighbor replication, path replication and multi-key replication. As argued in [55], the data replication strategy can have a significant impact on the system performance. Further effort may go into exploring more "dynamic" and adaptive replication approaches as function of data popularity or average peer online probability [52]. Although the replication techniques increase the data availability, they come not only at the expense of more storage space but can also affect the *data consistency* (e.g., concurrent update for the same triple). Moreover, the data inconsistency issue becomes even more intricate under the partitioning of the P2P network. Thus, an update operation might not address all replicas as a node storing a replica can be offline during the update process. Therefore, trade-offs are made between high data availability, data consistency and partition-tolerance. Brewer brought all these trade-offs together and presented the CAP theorem [40] which states that with Consistency, Availability, and Partition-tolerance, we can only ensure two out of these three properties. Recognizing which of the "CAP" properties the application really need is a fundamental step in building a successful distributed, scalable, highly reliable system.

Query processing. From the query processing point of view, sharing RDF data imposes new challenges on the distributed storage infrastructure related to supporting advanced query processing algorithms beyond simple keyword-based retrieval. Therefore, we need to take a deeper look at how the query can be optimized before being processed. The presented approaches for data retrieval and integration are mainly achieved either by fetching triples to query's originator which coordinates the query evaluation or forwarding intermediate results (e.g., [23, 59]) whenever the query is partially resolved. However, we believe that the first approach may not efficiently resolve conjunctive queries where each sub-query leads to a huge result set while the final join operation between them conducts to a small set. *Caching* results [14, 59] can alleviate this challenge, at least for similar queries, but can also affect the data consistency. Thereby, a trade-off is made between the network resource usage on one side and the information staleness on the other side. Other approaches such as in [14] decide at runtime whether the current result set has to be fetched to the query's originator or continue to be forwarded to other neighbors. As the query processing becomes more crucial especially when processing a huge data set such as the Billion Triple Challenge 2009 (BTC2009) dataset[3], some works have been proposed to reduce the large data sets to the interesting subsets as in [41]. To do so, BitMat [13], based on Bit Matrix conjunctive query execution approach, is used to generate compressed RDF structure. Therefore, a dominant challenge related to the distributed query processing [53] is how to improve the query performance and find an "optimal" query plan in order to enhance the query performance and reduce the communication cost. An already explored direction towards the query optimization plan is introduced by OptARQ [19], based on Jena ARQ [2], which uses the concept of *triple pattern selectivity estimation*. This concept makes use of statistical information about the ontological resources and evaluates the fraction of triples satisfying a given pattern. The goal is to find the query execution plan that minimizes the intermediate result set size of each triple pattern by join re-ordering strategies. Thereof, the smaller the selectivity the less intermediate results are likely to be produced. The presented works focusing on continuous queries [10, 24, 42, 54, 57, 58, 74], summarized in Table 3, have slightly different mechanisms for dealing with notifications and subscriptions. The only work based on a unstructured overlay network [10] relies on super-peers to manage and process subscriptions as well as route notifications while most of them are based on structured P2P overlays [24, 42, 54, 57, 58, 74] and directly take advantage of their underlying indexing mechanisms. Triples are indexed and retrieved using cryptographic functions (except for [74]). Ranger *et al.* [74] do not index data *a priori* but reuse the subscription management and group communication primitives used for notifications' propagation offered by Scribe [26]. By adding several caching mechanisms, in order to maintain the most popular queries and respective results fresh, they enhanced the data retrieval mechanism.

Pub/sub systems have been studied for some time now and are well established, but there exists a strong need for Quality of Service (QoS) [64], especially when deployed on top of a *best-effort* infrastructure such as the Internet. Now we will discuss some improvement opportunities with an emphasis on QoS. These improvements, even if they are valid for a lot of systems, were selected because

---

[3] http://vmlion25.deri.ie/

Table 3. Summary of the presented publish/subscribe systems for RDF data storage and retrieval

Pub/Sub System	Contribution highlights
Cai et al. [24]	– Subscription to atomic, disjunctive, range and conjunctive multi-predicate queries is possible. – Some combinations of conjunctive and disjunctive queries are not supported in the subscriptions. – A replication mechanism is provided for both subscriptions and data (for atomic queries), inherited from MAAN [22]. – Support for highly skewed subscription patterns is offered.
Chirita et al. [10]	– Super-peer-based system. – Formalism of the proposed system using a *typed first-order* language is given. – Resources advertisement managed by the super-peers using local partially order sets to reduce the network traffic. – Subscriptions are arranged in a hierarchical structure and take advantage of the *subscription subsumption* technique. – Notification matching and offline notifications' management are done by super-peers. – Integrated authentication mechanisms.

SQC and MQC [58]	– RDF queries in the style of RDQL.
	– Conjunctive multi-predicates queries are supported through two DHT-agnostic algorithms that extend the Query Chain (QC) algorithm [59]
	  • *Single Query Chain* (SQC)
	  • *Multiple Query Chain* (MQC)

CQC and CSBV [57]	– Only focuses on conjunctive queries (the motivation being that it is a core construct of RDF query languages).
	– Arbitrary continuous conjunctive queries are considered through two algorithms:
	  • *Continuous Query Chain* (CQC): multiple nodes, forming a chain of fixed length, participate in the resolution of a subscription.
	  • *Continuous Spread-By-Value* (CSBV): can be seen as a dynamic version of CQC where participating nodes are chosen during the subscription's processing.
	– Optimizations are provided to reduce network traffic, such as IP caching or query chain optimization.

Ranger et al. [74]	– Queries are expressed in a SPARQL dialect. – No data indexing *a priori* is used. – Various caching mechanisms (cached lookups, redundant caching, etc.) and grouping of related predicates of the same subject on the same node. – Queries are translated into a tree of atomic or complex queries. – Various logical trees of nodes are created during subscription processing, in which they can either *(i)* actively participate in a query resolution, *(ii)* forward query results to other nodes, *(iii)* simply act as consumers of query results.

few of presented pub/sub works actually focus on QoS. Enabling the Semantic Web at the Internet level means that we have no control on the architecture the application is deployed upon. Therefore, mechanisms dealing with this orthogonal issue should be taken into consideration.

Improvements related to subscription processing. Several techniques exist in order to improve the processing of subscriptions. Subscription *subsumption* [47] and *summarization* [85] are two optimization techniques whose goals are to reduce subscription dissemination traffic and enhance processing. The former exploits the "covering" relationship between a pair of subscriptions while the latter aims at representing them in a much more compact form. Well known pub/sub systems such as SIENA [25] and PADRES [56] implement *pair-wise* subscription cover checking to reduce redundancy in subscription forwarding. A more efficient approach, argued in [47], would be to exploit the subscription subsumption relationship between a new subscription and a set of existing active ones. In the presented pub/sub systems, only [10] takes advantage of this kind of subscription subsumption technique, while none uses summarization. We have yet to see the combination of these two improvement techniques applied to RDF-based pub/sub systems.

Improvements related to notification processing. Depending on the type of the application, there might be the need to ensure a (*stronger or weaker*) form of reliability as far as notifications are concerned. For instance, in a financial system such as the New York Stock Exchange (NYSE) or the the French Air Traffic Control System, reliability is *critical* as argued in [20]. Ensuring a correct dissemination of events despite failures requires a particular form of resiliency. As stated earlier, most of the presented works do not extensively offer QoS guarantees from the notification point of view (or even for subscription for that matter). Some of them use replication techniques to ensure data availability, but few if any discuss other aspects of QoS in depth such as the latency, bandwidth, delivery semantics, reliability and message ordering. As argued in [64], a lot of efforts are underway to build a generation of QoS-aware pub/sub systems. As a matter of fact, adaptation and QoS-awareness constitute major open research challenges that are naturally present in RDF-based pub/sub systems. The following three classes of service guarantees constitute some interesting challenges:

(i) Delivery semantics. Delivery semantics comes into play at the last hop, prior to notifying the subscriber. Notifications can be delivered *at most once, at least once, exactly once* or in a completely *best effort* way, that is with no guarantees of delivery whatsoever. All these delivery guarantees are implemented by a wide variety of broadcast algorithms which can be found in [43]. The message complexity in these algorithms is generally high and thus, depending also on the application guarantee requirements, they will have to be adapted to the overlay topology of the P2P system.

(ii) Reliability. The few works providing service guarantees generally follow a publisher-offered, subscriber (client)-requested pattern. None of the works take the reverse approach, that is, *providing application-specific quality of service guarantees explicitly specified by the client*, as argued in [63]. Earlier works at the overlay layer such as RON [11] and TAROM [84] focused their efforts on resiliency from the ground up, the problem is that they "only" consider link reliability without taking into account the node quality (e.g., load, subscribers/publishers degree, churn rate, etc.). Hermes [73] explicitly deals with routing robustness by introducing a reliability model at the routing level which enables event brokers to recover from failures. However, Hermes does not provide any client-specified service level reliability guarantees. Pub/sub systems which consider *event routing* based on reliability requirements are rare schemes as pointed out in [65].

(iii) Message ordering. In pub/sub systems, some applications, such as such as in financial systems and air traffic control emphasize on the preservation of a *temporal* order between the reception of events (e.g., FIFO/LIFO, priority, causal or total, etc.). Adopting a message ordering can have a significant impact not only on the routing algorithms but also on how the subscriber's node processes arriving messages. Strong dissemination abstractions such as a *Reliable Causal Broadcast*, well known and studied in the distributed systems literature, provide strong guarantees. However, because of their inherent inability to scale, their usage cost is prohibitive and limits their application in large-scale settings, as

argued in [43, 48]. Since the majority of the systems presented are built on top of structured overlay networks, one should take advantage of the overlay structure to disseminate events in a more efficient way, as in [26, 35]. Overall, there is a lot of room for improvements in pub/sub systems from a QoS point of view as pointed out, and the Semantic Web just might give us the opportunity to finally make these systems more reliable.

5   Conclusion

The Semantic Web enables users to model, manipulate and query information at a conceptual level in an unprecedented way. The underlying goal of this grand vision is to allow people to exchange information and link the conceptual layers of applications without falling into technicalities. At the core of the Semantic Web lies a powerful data model - RDF - an incarnation of the universal relation model developed in the 1980ies [66]. P2P technologies address *system complexity* by abandoning the centralization of knowledge and control in favor of a decentralized information storage and processing. The last generation of P2P networks - structured overlay networks [36] - represents an important step towards practical scalable systems. This survey presented various research efforts on RDF data indexing and retrieval in P2P systems. RDF-based P2P approaches discussed in this survey combine two research directions: research efforts that concentrate on the data model and query languages and efforts that attempt to take advantage of the expressiveness and the flexibility of such data model. The main objective was to build large scale distributed applications using the P2P technologies in order to come up with fully distributed and scalable infrastructures for RDF data storage and retrieval. As a natural extension, the publish/subscribe paradigm, built atop a P2P substrate, illustrates a more dynamic way of querying RDF data in a continuous manner. The main goal of these systems is to offer advanced query mechanisms and sophisticated information propagation among interested peers. Even though they are built on top of P2P systems, thus encompassing already discussed research directions found at the P2P level, they incorporate research issues of their own in terms of subscription processing and notification propagation. Leveraging orthogonal ideas, such as *federation*, could provide further control on the overall scalability and autonomy of the system by allowing different entities to keep the control on their own data while still collaborating. This structure inspired the work proposed by Baude *et al.* [17] to create the *Semantic Spaces*.

We have seen that the combination of P2P technologies such as structured overlay networks and the RDF data model is a promising step towards an effective Internet-scale Semantic Web. Both research communities made a tremendous effort, both on their own side, to provide solid and sound foundations. This interdisciplinary effort has proven that both communities can benefit from one another and that they can push forward the existing Web frontier.

References

[1] Gnutella RFC, http://rfc-gnutella.sourceforge.net/
[2] Jena - a Semantic Web Framework for java, http://jena.sourceforge.net/
[3] RDFStore, http://rdfstore.sourceforge.net/
[4] Resource Description Framework, http://www.w3.org/RDF/
[5] SETI@home, http://setiathome.ssl.berkeley.edu/
[6] SPARQL Query Language, http://www.w3.org/TR/rdf-sparql-query/
[7] W3C Semantic Web Activity, http://www.w3.org/2001/sw/
[8] Aberer, K., Cudré-Mauroux, P., Datta, A., Despotovic, Z., Hauswirth, M., Punceva, M., Schmidt, R.: P-Grid: a self-organizing structured P2P system. SIG-MOD Record 32(3), 33 (2003)
[9] Aberer, K., Cudré-Mauroux, P., Hauswirth, M., Pelt, T.V.: GridVine: Building Internet-Scale Semantic Overlay Networks. In: International Semantic Web Conference (2004)
[10] Alex, P., Chirita, R., Idreos, S., Koubarakis, M., Nejdl, W.: Designing semantic publish/subscribe networks using super-peers. In: Semantic Web and Peer-To-Peer (January 2004)
[11] Andersen, D., Balakrishnan, H., Kaashoek, F., Morris, R.: Resilient overlay networks. In: Proceedings of the 18th ACM Symposium on Operating Systems Principles, pp. 131–145. ACM, Banff (2001)
[12] Antoniou, G., Harmelen, F.: Web ontology language: Owl. In: Handbook on Ontologies, pp. 91–110 (2009)
[13] Atre, M., Srinivasan, J., Hendler, J.: BitMat: A Main-memory Bit Matrix of RDF Triples for Conjunctive Triple Pattern Queries. In: 7th International Semantic Web Conference (ISWC) (October 2008)
[14] Battre, D.: Caching of intermediate results in DHT based RDF stores. Int. J. Metadata Semant. Ontologies 3(1), 84–93 (2008)
[15] Battré, D.: Query Planning in DHT Based RDF Stores. In: Proceedings of the 2008 IEEE International Conference on Signal Image Technology and Internet Based Systems (SITIS), pp. 187–194. IEEE Computer Society, Washington, DC, USA (2008)
[16] Battré, D., Heine, F., Höing, A., Kao, O.: On triple dissemination, forward-chaining, and load balancing in DHT based RDF stores. In: Moro, G., Bergamaschi, S., Joseph, S., Morin, J.-H., Ouksel, A.M. (eds.) DBISP2P 2005 and DBISP2P 2006. LNCS, vol. 4125, pp. 343–354. Springer, Heidelberg (2007)
[17] Baude, F., Filali, I., Huet, F., Legrand, V., Mathias, E., Merle, P., Ruz, C., Krummenacher, R., Simperl, E., Hamerling, C., Lorré, J.-P.: ESB Federation for Large-Scale SOA. In: Proceedings of the ACM Symposium on Applied Computing (SAC), pp. 2459–2466 (2010)
[18] Berners-Lee, T.: Linked data. W3C Design Issues (2006)
[19] Bernstein, A., Kiefer, C., Stocker, M.: OptARQ: A SPARQL Optimization Approach based on Triple Pattern Selectivity Estimation. Tech. rep., University of Zurich (2007)
[20] Birman, K.P.: A review of experiences with reliable multicast. Software: Practice and Experience 29(9), 741–774 (1999)
[21] Bloom, B.H.: Space/Time Trade-offs in Hash Coding With Allowable Errors. Commun. ACM 13(7), 422–426 (1970)
[22] Cai, M., Frank, M., Chen, J., Szekely, P.: MAAN: A multi-attribute Addressable Network for Grid Information Services. Journal of Grid Computing 2 (2003)

[23] Cai, M., Frank, M.R.: RDFPeers: a scalable distributed RDF Repository Based on a Structured Peer-to-Peer Network. In: WWW, pp. 650–657 (2004)

[24] Cai, M., Frank, M.R., Yan, B., MacGregor, R.M.: A subscribable Peer-to-Peer RDF Repository for Distributed Metadata Management. J. Web Sem. 2(2), 109–130 (2004)

[25] Carzaniga, A., Rosenblum, D.S., Wolf, A.L.: Design and Evaluation of a Wide-Area Event Notification Service. ACM Trans. Comput. Syst. 19(3), 332–383 (2001)

[26] Castro, M., Druschel, P., Kermarrec, A., Rowstron, A.: SCRIBE: A large-scale and decentralized application-level multicast infrastructure. IEEE Journal on Selected Areas in Communications 20(8), 1489–1499 (2002)

[27] Castro, M., Costa, M., Rowstron, A.: Debunking some myths about structured and unstructured overlays. In: Proceedings of the 2nd Conference on Symposium on Networked Systems Design and Implementation (NSDI), pp. 85–98. USENIX Association (2005)

[28] Magiridou, M., Sahtouris, S., Christophides, V., Koubarakis, M.: RUL: A declarative update language for RDF. In: Gil, Y., Motta, E., Benjamins, V.R., Musen, M.A. (eds.) ISWC 2005. LNCS, vol. 3729, pp. 506–521. Springer, Heidelberg (2005)

[29] Cohen, B., http://www.bittorrent.com

[30] Crespo, A., Garcia-Molina, H.: Semantic Overlay networks for P2P Systems. In: Agents and Peer-to-Peer Computing, pp. 1–13 (2005)

[31] Cudré-Mauroux, P., Agarwal, S., Aberer, K.: Gridvine: An infrastructure for peer information management. IEEE Internet Computing 11, 36–44 (2007)

[32] Dan Brickley, R.G.: RDF Vocabulary Description Language 1.0: RDF schema, http://www.w3.org/TR/rdf-schema/

[33] Della Valle, E., Turati, A., Ghioni, A.: *PAGE*: A distributed infrastructure for fostering RDF-based interoperability. In: Eliassen, F., Montresor, A. (eds.) DAIS 2006. LNCS, vol. 4025, pp. 347–353. Springer, Heidelberg (2006)

[34] Demers, A., Greene, D., Hauser, C., Irish, W., Larson, J., Shenker, S., Sturgis, H., Swinehart, D., Terry, D.: Epidemic algorithms for replicated database maintenance. In: Proceedings of the Sixth Annual ACM Symposium on Principles of Distributed Computing, pp. 1–12. ACM, Vancouver (1987)

[35] El-Ansary, S., Alima, L., Brand, P., Haridi, S.: Efficient broadcast in structured P2P networks. In: Peer-to-Peer Systems II, pp. 304–314. Springer, Heidelberg (2003)

[36] El-Ansary, S., Haridi, S.: An overview of structured overlay networks. In: Theoretical and Algorithmic Aspects of Sensor, Ad Hoc Wireless and Peer-to-Peer Networks. CRC Press, Boca Raton (2005)

[37] Eugster, P., Guerraoui, R., Kermarrec, A.M., Massoulie, L.: From Epidemics to Distributed Computing. IEEE Computer 37(5), 60–67 (2004)

[38] Eugster, P., Felber, P., Guerraoui, R., Kermarrec, A.: The many faces of publish/subscribe. ACM Computing Surveys (CSUR) 35(2), 114–131 (2003)

[39] Filali, I., Bongiovanni, F., Huet, F., Baude, F.: RDF Data Indexing and Retrieval: A survey of Peer-to-Peer based solutions. Research Report RR-7457, INRIA (November 2010)

[40] Gilbert, S., Lynch, N.: Brewer's Conjecture and the Feasibility of Consistent Available Partition-Tolerant Web Services. In: ACM SIGACT News, p. 2002 (2002)

[41] Williams, G.T., Weaver, J., Atre, M., Hendler, J.A.: Scalable Reduction of Large Datasets to Interesting Subsets. In: 8th International Semantic Web Conference (2009)

[42] Gu, T., Pung, H.K., Zhang, D.: Information Retrieval in Schema-based P2P Systems Using One-dimensional Semantic Space. Computer Networks 51(16), 4543–4560 (2007)

[43] Guerraoui, R., Rodrigues, L.: Introduction to reliable distributed programming. Springer-Verlag New York Inc., Secaucus (2006)

[44] Harth, A., Decker, S.: Optimized Index Structures for Querying RDF from the Web. In: Proceedings of the Third Latin American Web Congress (LA-WEB), p. 71. IEEE Computer Society, Washington, DC, USA (2005)

[45] Heine, F.: Scalable p2p based RDF querying. In: Proceedings of the 1st International Conference on Scalable Information Systems (InfoScale), p. 17. ACM, New York (2006)

[46] Broekstra, J.: Ehrig and Peter Haase and Frank van Harmelen and Maarten Menken and Peter Mika and Bjorn Schnizler and Ronny Siebes: Bibster - A Semantics-Based Bibliographic Peer-to-Peer System. In: The Second Workshop on Semantics in Peer-to-Peer and Grid Computing (SEMPGRID), New York (May 2004)

[47] Jafarpour, H., Hore, B., Mehrotra, S., Venkatasubramanian, N.: Subscription subsumption evaluation for content-based publish/Subscribe systems. In: Issarny, V., Schantz, R. (eds.) Middleware 2008. LNCS, vol. 5346, pp. 62–81. Springer, Heidelberg (2008)

[48] de Juan, R., Decker, H., Miedes, E., Armendariz, J.E., Munoz, F.D.: A Survey of Scalability Approaches for Reliable Causal Broadcasts. Tech. Rep. ITI-SIDI-2009/010 (2009)

[49] Karnstedt, M., Sattler, K.U., Richtarsky, M., Muller, J., Hauswirth, M., Schmidt, R., John, R., Ilmenau, T.U.: UniStore: querying a DHT-based universal storage. In: IEEE 23rd International Conference on Data Engineering (ICDE), pp. 1503–1504 (2007)

[50] Karvounarakis, G., Alexaki, S., Christophides, V., Plexousakis, D., Vouton, F.V., Scholl, M.: RQL: A Declarative Query Language for RDF. In: Proceedings of the 11th International Conference on World Wide Web (WWW), pp. 592–603. ACM Press, New York (2002)

[51] King, R.A., Hameurlain, A., Morvan, F.: Query Routing and Processing in Peer-To-Peer Data Sharing Systems. International Journal of Database Management Systems, 116–139 (2010)

[52] Knežević, P., Wombacher, A., Risse, T.: DHT-Based Self-adapting Replication Protocol for Achieving High Data Availability. In: Advanced Internet Based Systems and Applications, pp. 201–210 (2009)

[53] Kossmann, D.: The State of The Art in Distributed Query Processing. ACM Comput. Surv. 32(4), 422–469 (2000)

[54] Koubarakis, M., Miliaraki, I., Kaoudi, Z., Magiridou, M., Papadakis-Pesaresi, A.: Semantic Grid Resource Discovery using DHTs in Atlas. In: Proceedings of 3rd GGF Semantic Grid Workshop, Athens, Greece (February 2006)

[55] Ktari, S., Zoubert, M., Hecker, A., Labiod, H.: Performance Evaluation of Replication Strategies in DHTs Under Churn. In: Proceedings of the 6th International Conference on Mobile and Ubiquitous Multimedia (MUM), pp. 90–97. ACM, New York (2007)

[56] Li, G., Hou, S., Jacobsen, H.: A Unified Approach to Routing, Covering and Merging in Publish/Subscribe Systems Based on Modified Binary Decision Diagrams. In: Proceedings of 25th IEEE International Conference on Distributed Computing Systems (ICDCS) 2005, pp. 447–457 (2005)

[57] Liarou, E., Idreos, S., Koubarakis, M.: Continuous RDF query processing over dHTs. In: Aberer, K., Choi, K.-S., Noy, N., Allemang, D., Lee, K.-I., Nixon, L.J.B., Golbeck, J., Mika, P., Maynard, D., Mizoguchi, R., Schreiber, G., Cudré-Mauroux, P. (eds.) ASWC 2007 and ISWC 2007. LNCS, vol. 4825, pp. 324–339. Springer, Heidelberg (2007)

[58] Liarou, E., Idreos, S., Koubarakis, M.: Publish/Subscribe with RDF data over large structured overlay networks. In: Moro, G., Bergamaschi, S., Joseph, S., Morin, J.-H., Ouksel, A.M. (eds.) DBISP2P 2005 and DBISP2P 2006. LNCS, vol. 4125, pp. 135–146. Springer, Heidelberg (2007)

[59] Liarou, E., Idreos, S., Koubarakis, M.: Evaluating conjunctive triple pattern queries over large structured overlay networks. In: Cruz, I., Decker, S., Allemang, D., Preist, C., Schwabe, D., Mika, P., Uschold, M., Aroyo, L.M. (eds.) ISWC 2006. LNCS, vol. 4273, pp. 399–413. Springer, Heidelberg (2006)

[60] Liu, Y., Plale, B.: Survey of Publish Subscribe Event Systems. Tech. rep., Indiana University (2003)

[61] Loo, B.T., Huebsch, R., Stoica, I., Hellerstein, J.M.: The Case for a Hybrid P2P Search Infrastructure. In: Peer-to-Peer Systems III, pp. 141–150 (2005)

[62] Lua, K., Crowcroft, J., Pias, M., Sharma, R., Lim, S.: A Survey and Comparison of Peer-to-Peer Overlay Network Schemes. IEEE Communications Surveys and Tutorials, 72–93 (2005)

[63] Mahambre, S.P., Bellur, U.: An Adaptive Approach for Ensuring Reliability in Event Based Middleware. In: Proceedings of the Second International Conference on Distributed Event-based Systems, pp. 157–168. ACM, New York (2008)

[64] Mahambre, S.P., Madhu Kumar, S.D., Bellur, U.: A Taxonomy of QoS-Aware, Adaptive Event-Dissemination Middleware. IEEE Internet Computing 11(4), 35–44 (2007)

[65] Mahambre, S., Bellur, U.: Reliable Routing of Event Notifications over P2P Overlay Routing Substrate in Event Based Middleware. In: IEEE International Parallel and Distributed Processing Symposium (IPDPS), pp. 1–8 (2007)

[66] Maier, D., Ullman, J.D., Vardi, M.Y.: On the Foundations of the Universal Relation Model. ACM Trans. Database Syst. 9(2), 283–308 (1984)

[67] Matono, A., Pahlevi, S., Kojima, I.: RDFCube: A P2P-Based Three-Dimensional Index for Structural Joins on Distributed Triple Stores. In: Databases, Information Systems, and Peer-to-Peer Computing, pp. 323–330 (2007)

[68] Meshkova, E., Riihijärvi, J., Petrova, M., Mähönen, P.: A Survey on Resource Discovery Mechanisms, Peer-to-Peer and Service Discovery Frameworks. Comput. Netw. 52(11), 2097–2128 (2008)

[69] Mu, Y., Yu, C., Ma, T., Zhang, C., Zheng, W., Zhang, X.: Dynamic Load Balancing With Multiple Hash Functions in Structured P2P Systems. In: Proceedings of the 5th International Conference on Wireless Communications, Networking and Mobile Computing (WiCOM), pp. 5364–5367. IEEE Press, Piscataway (2009)

[70] Mühl, G., Fiege, L., Pietzuch, P.: Distributed Event-Based Systems. Springer-Verlag New York, Inc., Secaucus (2006)

[71] Nejdl, W., Wolf, B., Qu, C., Decker, S., Sintek, M., Naeve, A., Nilsson, M., Palmer, M., Risch, T.: Edutella: A P2P Networking Infrastructure Based on RDF. In: Proceedings of the 11 International World Wide Web Conference, WWW (May 2002)

[72] Nejdl, W., Wolpers, M., Siberski, W., Schmitz, C., Schlosser, M., Brunkhorst, I., Löser, A.: Super-peer-based routing and clustering strategies for RDF-based peer-to-peer networks. In: Proceedings of the 12th International Conference on World Wide Web, pp. 536–543. ACM, New York (2003)

[73] Pietzuch, P., Bacon, J.: Hermes: A distributed Event-based Middleware Architecture. In: Proceedings of the 22nd International Conference on Distributed Computing Systems(ICDCS), pp. 611–618 (2002)

[74] Ranger, D., Cloutier, J.F.: Scalable Peer-to-Peer RDF Query Algorithm. In: Proceedings of Web information systems engineering International Workshops (WISE), p. 266 (November 2005)

[75] Ratnasamy, S., Francis, P., Handley, M., Karp, R., Shenker, S.: A Scalable Content-Addressable Network. In: Proceedings of the 2001 Conference on Applications, Technologies, Architectures, and Protocols for Computer Communications (SIGCOMM), pp. 161–172. ACM, New York (2001)

[76] Rhea, S., Geels, D., Roscoe, T., Kubiatowicz, J.: Handling Churn in a DHT. In: Proceedings of the Annual Conference on USENIX Annual Technical Conference (ATEC), p. 10 (2004)

[77] Risson, J., Moors, T.: Survey of Research Towards Robust Peer-to-Peer Networks: Search Methods. Computer Networks 50(17), 3485–3521 (2006)

[78] Rowstron, A., Druschel, P.: Pastry: Scalable, decentralized object location, and routing for large-scale peer-to-peer systems. In: Liu, H. (ed.) Middleware 2001. LNCS, vol. 2218, pp. 329–350. Springer, Heidelberg (2001)

[79] Guha, R.V.: rdfDB: An RDF Database, http://guha.com/rdfdb/

[80] Schlosser, M.T., Sintek, M., Decker, S., Nejdl, W.: HyperCuP - Hypercubes, Ontologies, and Efficient Search on Peer-to-Peer Networks. In: Moro, G., Koubarakis, M. (eds.) AP2PC 2002. LNCS (LNAI), vol. 2530, pp. 112–124. Springer, Heidelberg (2003)

[81] Seaborne, A.: RDQL - A Query Language for RDF. Tech. rep., W3C, proposal (2004)

[82] Staab, S., Stuckenschmidt, H.: Semantic Web and Peer-to-Peer. Springer, Heidelberg (2006)

[83] Stoica, I., Morris, R., Karger, D., Kaashoek, M.F., Balakrishnan, H.: Chord: A Scalable Peer-to-Peer Lookup Service for Internet Applications. In: Proceedings of the 2001 Conference on Applications, Technologies, Architectures, and Protocols for Computer Communications (SIGCOMM), pp. 149–160. ACM, New York (2001)

[84] Tang, C., McKinley, P.: Improving Multipath Reliability in Topology-Aware Overlay Networks. In: proceedings of the 25th IEEE International Conference on Distributed Computing Systems Workshops (ICDCSW), pp. 82–88. IEEE, Los Alamitos (2005)

[85] Triantafillou, P., Economides, A.: Subscription Summarization: A New Paradigm for Efficient Publish/Subscribe Systems. In: Proceedings of the 24th International Conference on Distributed Computing Systems (ICDCS 2004), pp. 562–571. IEEE Computer Society, Los Alamitos (2004)

[86] Voulgaris, S., Rivière, E., Kermarrec, A.M., Steen, M.V.: Sub-2-sub: Self-organizing content-based publish subscribe for dynamic large scale collaborative networks. In: Proceedings of the fifth International Workshop on Peer-to-Peer Systems, IPTPS (2006)

[87] Yang, B., Garcia-Molina, H.: Designing a Super-Peer Network. In: Proceedings of the 19th International Conference on Data Engineering (ICDE), vol. 1063, p. 17 (2003)

[88] Zhou, J., Hall, W., Roure, D.D.: Building a Distributed Infrastructure for Scalable Triple Stores. Journal of Computer Science and Technology 24(3), 447–462 (2009)

A Semantic-Based Approach for Data Management in a P2P System

Damires Souza[1], Carlos Eduardo Pires[2], Zoubida Kedad[3],
Patricia Tedesco[4], and Ana Carolina Salgado[4]

[1] Federal Institute of Education, Science and Technology of Paraiba (IFPB), Brazil
damires@ifpb.edu.br
[2] Federal University of Campina Grande (UFCG), Computer Science Department
Av. Aprígio Veloso, 882, Bodocongó - 58109-970 - Campina Grande, PB, Brazil
cesp@dsc.ufcg.edu.br
[3] Université de Versailles et Saint-Quentin-en-Yvelines (UVSQ)
45 Avenue des Etats-Unis, 78035 Versailles, France
zoubida.kedad@prism.uvsq.fr
[4] Federal University of Pernambuco (UFPE), Center for Informatics
P.O. Box 7851, 50.732-970, Recife, PE, Brazil
{pcart,acs}@cin.ufpe.br

Abstract. Data management in P2P Systems is a challenging problem, due to the high number of autonomous and heterogeneous peers. In some Peer Data Management Systems (PDMSs), peers are semantically clustered in the overlay network. A peer joining the system is assigned to an appropriate cluster, and a query issued by a user at a given peer is routed to semantic neighbor clusters which can provide relevant answers. To help matters, semantic knowledge in the form of ontologies and contextual information has been used successfully to support the techniques used to manage data in such systems. Ontologies can be used to solve the heterogeneities between the peers, while contextual information allows a PDMS to deal with information that is acquired dynamically during the execution of a given query. The goal of this paper is to point out how the semantics provided by ontologies and contextual information can be used to enhance the results of two important data management issues in PDMSs, namely, peer clustering and query reformulation. We present a semantic-based approach to support these processes and we report some experimental results which show how semantics can improve them.

Keywords: PDMS, Semantics, Data Management, Ontologies, Context, Clustering.

1 Introduction

Peer-to-Peer (P2P) systems are massively distributed and highly volatile systems for sharing large amounts of resources. Peers communicate through a self-organizing and fault tolerant network topology which runs as an overlay on top of the physical network. Peer Data Management Systems (PDMSs) [1, 2, 3, 4, 5, 6] came into the focus

A. Hameurlain, J. Küng, and R. Wagner (Eds.): TLDKS III , LNCS 6790, pp. 56–86, 2011.

of research as a natural extension to distributed databases in the P2P setting. PDMSs are P2P applications where each peer represents an autonomous data source which exports either its entire data schema or only a portion of it. Such schemas, named exported schemas, represent the data to be shared with the other peers. To enable data sharing, correspondences between elements belonging to these exported schemas are generated and maintained.

In general, PDMSs are rather useful for query answering, when people have overlapping data and they want to access additional related information stored in other peers [6]. For example, different universities can be interested in sharing data about their research projects as well as in producing more complete results that involve the overall distributed data in the university peers.

Data management in PDMSs is a challenging problem given the large number of peers, their autonomous nature, and the heterogeneity of their schemas. To help matters, semantics has shown to be a helpful support for the techniques used for managing data (e.g., query answering or schema matching). In this work, semantics concerns the task of assigning meaning to schema elements [4, 7, 8, 9, 10] or expressions that need to be interpreted in a given situation. In distributed environments such as PDMSs, the use of semantics is supported by background knowledge such as domain ontologies as well as by information that is acquired statically or on the fly such as users' preferences or peers availability.

In our setting, semantic knowledge is mainly obtained through ontologies and contextual information. We use ontologies in a threefold manner: (i) as a standard model to represent peer's metadata; (ii) as a mechanism to represent and store contextual information; and (iii) as background knowledge to identify semantic correspondences between matching ontologies which represent peers' schemas. For instance, in (iii) we use logical disjointness axioms present in a domain ontology to infer that two concepts from different matching ontologies are disjoint as well.

We define *Context* as a set of elements surrounding a domain entity of interest (e.g., user, query, and peer) which is considered relevant in a specific situation during some time interval [11, 12, 13, 14]. We use context as a kind of semantic knowledge that enhances the overall query reformulation process by assisting it to deal with information that can only be acquired on the fly (e.g., according to the availability of data sources, which is perceived at run time, specific query routing strategies can be executed). In this light, the main goal of this paper is to show how the semantics provided by ontologies and contextual information can be used to enhance data management issues in PDMSs. To this end, we present a PDMS, named SPEED, which adopts a semantic-based approach to assist in solving relevant issues in data management.

There are a number of relevant issues concerned with the use of semantics in PDMS. In particular, in a PDMS scenario, a key challenge is query answering, where queries submitted at a peer are answered with data residing at that peer and with data acquired from neighbor peers through the use of correspondences. An important step in this task is the reformulation of a query posed at a peer into a new query expressed in terms of a target peer. Another important point is that query answering in PDMS can be improved if semantically similar peers are put together in the overlay network. Considering that, the main categories of problems which have been initially addressed by this work are peer clustering and query reformulation.

In summary, our main contributions are: (i) a PDMS architecture mainly designed to group semantic similar peers within clusters; (ii) an incremental clustering process considering the dynamic aspect of P2P networks; (iii) a query reformulation algorithm considering new semantic inter-schema correspondences (e.g. closeness and disjointness) and contextual information; and (iv) experimentations of the two main proposed processes: peer clustering and query reformulation.

The paper is organized as follows: Section 2 describes the architecture of SPEED. Section 3 presents the semantic-based process for peer clustering in SPEED. Section 4 describes the semantic-based query reformulation approach used in SPEED. Experiments and results are presented in Section 5. Related work is discussed in Section 6. Finally, Section 7 draws our conclusions and points out some future work.

2   A PDMS Architecture

Our PDMS, named *SPEED* (**S**emantic **PEE**r-to-Peer **D**ata Management System) [15] uses a mixed P2P network topology: unstructured [16], structured [17], and super-peer [18]. In an unstructured network, peers may join and leave the network without any notification and may connect to whomever they wish. In a structured network, peers are organized in a rigid structure and connections between peers are fixed according to a certain protocol. A super-peer network exploits the heterogeneity among participating peers, by assigning greater responsibility to those super-peers who have high network bandwidth, large storage capacity, and significant processing power. The strengths of such topologies are explored to assist peer organization in the overlay network according to their exported schemas. SPEED's main goal is to cluster semantically similar peers to ease the establishment of semantic correspondences between peers and, consequently, to improve query answering on a large number of relevant data sources [19]. Next, we present an overview of SPEED's architecture.

2.1   Architecture Overview

As illustrated in Figure 1, SPEED has three distinct types of peers: *data peers, integration peers*, and *semantic peers*. A **data peer** is a simple computer or a server storing an autonomous data source. DP_{ij1}, DP_{ij2}, and DP_{ijk} are examples of data peers. Data peers are logically organized in a super-peer network. Thus, **semantic clusters** (clusters, in short) are formed according to data peers' exported schema. Each cluster has a special type of peer named **integration peer**. In fact, an integration peer is a data peer with higher computational capacity. It is responsible for defining and maintaining schema correspondences as well as for managing query answering and data peer's metadata. For instance, IP_{ij} is the integration peer of the cluster formed by the data peers DP_{ij1}, DP_{ij2}, and DP_{ijk}.

A logical set of clusters sharing content belonging to a common knowledge domain forms **semantic community** (community, in short). Each community has a special type of peer named **semantic peer**. SP_i is an example of a semantic peer. Semantic peers are connected through a structured network, while integration peers are connected through an unstructured network. A semantic peer acts as an entry point for its community. It is responsible for forwarding a requesting peer to an initial cluster CL_{ij}.

This initial cluster is obtained from a **semantic index**. A semantic index is a data structure located at a semantic peer summarizing the content available in the clusters of a community. Each index entry stores a summary of the peers' schemas in a particular cluster and the network address of the corresponding integration peer. In the following, we explain how the semantic knowledge expressed by ontologies is used as a way to enrich some important services in SPEED.

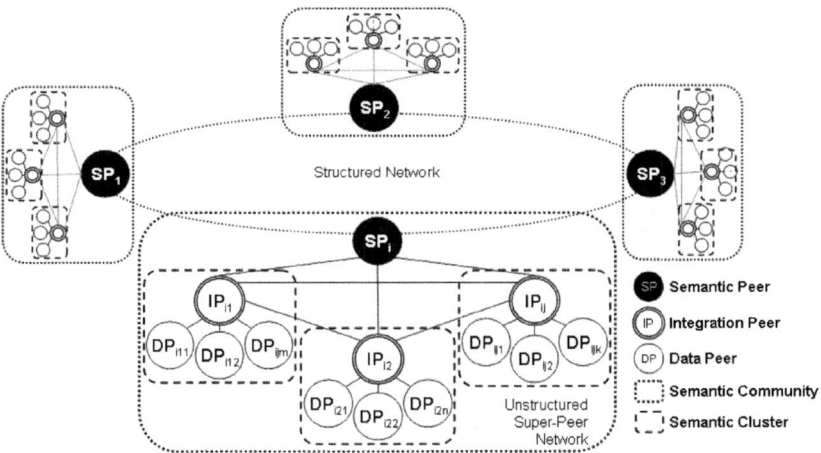

Fig. 1. An overview of the SPEED's architecture

2.2 Ontologies in SPEED

In SPEED, four distinct types of ontologies are employed (Figure 2): (i) *local ontologies*, representing the schema of the sources stored in data peers and integration peers; (ii) *cluster ontologies*, which are obtained by merging the local ontologies of the cluster data peers; (iii) *summarized cluster ontologies*, representing a cluster ontology succinctly; and (iv) *community ontologies* (i.e., domain ontologies), containing concepts and properties of a particular knowledge domain.

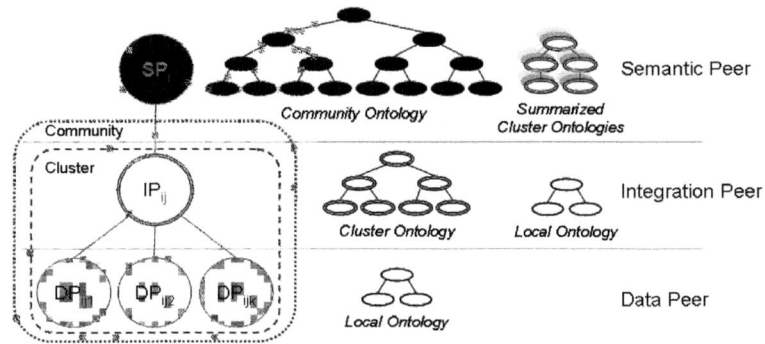

Fig. 2. The different types of ontologies used in SPEED

Each peer contains one or more internal modules whose main goal is to deal with the different types of ontologies. For instance, an integration peer contains internal modules which are responsible for matching and merging ontologies. A semantic peer has a module to summarize cluster ontologies to be used as a semantic index. A data peer has a module to translate an exported schema described in its original data model to an ontology representation. To this end, an ontology matching tool (called *SemMatcher* [20]) and an ontology summarization tool (called *OWLSum* [21]) have been developed. Besides, we have also implemented a simple merge module, denoted *OntMerger* [22] that takes as arguments two peer ontologies (i.e., a cluster ontology and a local ontology) and the set of correspondences between them (generated by *SemMatcher*). Due to their relevance to our work, in the following sections, we provide a running example and describe the *SemMatcher* and *OWLSum* tools.

2.3 Running Example

Considering the SPEED setting depicted in Figure 1, our running example scenario is composed by two integration peers P_1 and P_2 which belong to the "Education" knowledge domain. In this current scenario, peers have complementary data about academic people and works (e.g., research) from different institutions. Each peer is provided with ontologies (O_1 and O_2) describing their schemas. Ontologies O_1 and O_2 are in OWL and are summarily depicted in Figure 3. Such views have been produced using OWLViz, a Protégé plug-in[1]. Sub-concepts are associated with their super-concepts through the *is-a* relationship. In addition, we have considered as a semantic reference a domain ontology concerning *Education*. Such ontology, named *UnivCSCMO.owl* is presented in Figure 4[2].

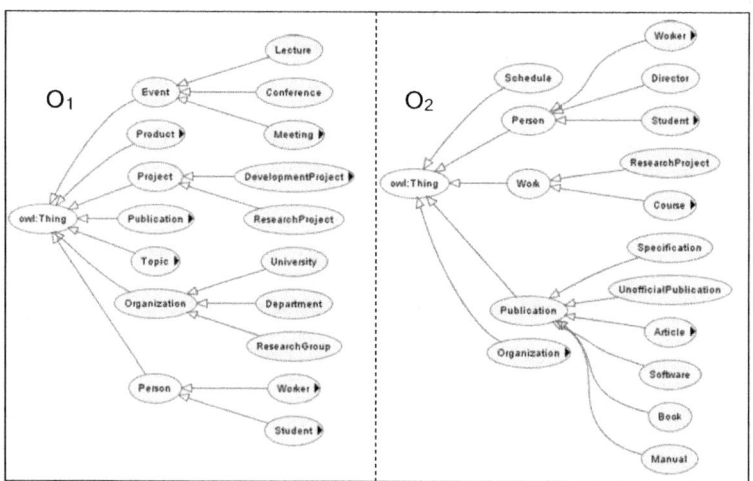

Fig. 3. Excerpts from Ontologies O_1 and O_2

---

[1] http://www.co-ode.org/downloads/owlviz/

[2] The complete ontologies are available at http://www.cin.ufpe.br/~speed/ontologies/Ontologies.html

Our semantic-based approach is able to deal with a degree of flexibility capable of accommodating a variety of scenarios. As presented in this running scenario, peer ontologies may have different levels of granularity, differing in terms of size, partition of concepts or conceptual organization. Throughout the paper, we will use the same ontologies presented in this section, discussing relevant aspects and showing how we match and use them in order to reformulate queries.

Fig. 4. Excerpt from the Education domain ontology (i.e., *UnivCSCMO.owl*)

2.4 The *SemMatcher* Tool

The process for matching peer ontologies used in *SemMatcher* brings together a combination of already defined matching strategies [23]. In this process, a linguistic-structural matcher and a semantic matcher are executed in parallel. The obtained similarity values of both matchers are combined through a weighted average. Each matcher receives a particular weight according to its importance in the matching process. As shown in Figure 5, the process receives as input two ontologies (O_1 and O_2) and a domain ontology DO to be used as background knowledge [20]. For instance, O_1 can be a cluster ontology, O_2 a local ontology, and DO a community ontology. As output, a set of correspondences (called A_{FI}) and a global similarity measure between O_1 and O_2 [20] are produced. The main steps carried out by the semantic-based ontology matching process are:

(1) *Linguistic-Structural Matching:* in this step, any existing ontology matching tool including linguistic and/or structural matchers can be used, e.g. H-Match [24].
(2) *Semantic Matching:* the semantic matcher identifies, besides the traditional types of correspondences (equivalence and subsumption), other ones such as closeness and disjointness [8]. Each generated semantic correspondence is organized considering a particular semantic ranking according to its level of confidence. In this light, we have initially assigned some weights to them, as follows: *isEquivalentTo* (1.0),

isSubConceptOf (0.8), *isSuperConceptOf* (0.8), *isCloseTo* (0.7), *isPartOf* (0.5), *isWholeOf* (0.5), and *isDisjointWith* (0.0). The weights reflect the degree of closeness between the correspondence elements, from the strongest relationship (equivalence) to the weakest one (disjointness).

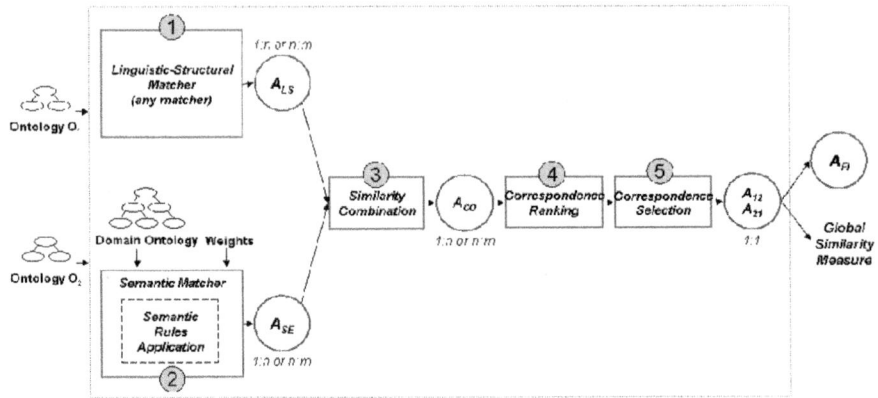

Fig. 5. The overall process for matching peer ontologies

(3) Similarity Combination: the individual similarity values of the correspondences produced by the linguistic-structural matcher and the semantic matcher are associated in a combined similarity value, through a weighted average of the values generated by the individual matchers.

(4) Correspondence Ranking: correspondences containing the elements of O_1 are ranked (in descending order) according to the associated similarity values.

(5) Correspondence Selection: a filtering strategy is applied to choose the most suitable correspondence for each O_1 element.

Steps 4 and 5 are also executed in the opposite direction. Correspondences containing the elements of O_2 are ranked according to the associated similarity values and the same filter strategy is applied. In order to evaluate the global similarity degree between O_1 and O_2, we take as input the sets of correspondences and apply existing similarity measures such as *dice* [26], *weighted* [27] and *overlap* [28].

2.5 The *OWLSum* Tool

As more data peers join a cluster, its corresponding cluster ontology can become larger and larger. Ontology summaries are used to minimize computation efforts when dealing with large-scale ontologies. OWLSum [21] is an ontology summarization tool that takes as argument a cluster ontology CLO_{ij} and automatically produces an abridged version of it (denoted OS_{ij}). The relevant concepts of CLO_{ij} are initially identified and OS_{ij} denotes the subontology of CLO_{ij} concentrating the maximum number of relevant concepts. Since relevant concepts can be non-adjacent in CLO_{ij}, non-relevant concepts may be also introduced in an OS_{ij}. OS_{ij} also corresponds to the subontology of CLO_{ij} containing the minimum number of non-relevant concepts.

The relevance of an ontology concept c_n is measured considering the relationships of c_n with other concepts in the cluster ontology CLO_{ij} (*centrality*) and the occurrences of c_n in the local ontologies $LO_1,\ldots,LO_n$ that compose CLO_{ij} (*frequency*). In our approach, centrality is used to capture the importance of a given concept within an ontology, whilst frequency is used when an ontology results from a merging process and captures the number of occurrences of a concept in the set of underlying local ontologies.

The main steps of the ontology summarization process are:

(1) *Calculate the relevance of each concept:* centrality and frequency are combined using a weighted formula in which the weights are defined according to the importance of each measure.

(2) *Determine the set of relevant concepts:* identifies the set of relevant concepts (denoted RC) of a cluster ontology CLO_{ij}. Ideally, the concepts in the identified set should be contained in the ontology summary OS_{ij}.

(3) *Group adjacent relevant concepts:* consists in forming groups of concepts containing only relevant concepts which are adjacent in the input cluster ontology CLO_{ij}. Three situations can occur: (i) each group is formed by a single relevant concept (all relevant concepts are non-adjacent in CLO_{ij}); (ii) at least one of the groups has more than one relevant concept (some relevant concepts are not adjacent in CLO_{ij}); and (iii) only one group is formed, containing all the relevant concepts. In the first two situations, the summarization process proceeds with Steps 4, 5 and 6. In the last situation, the summarization process finishes and the ontology summary corresponds to the identified group of concepts.

(4) *Identify paths between groups of relevant concepts:* detects all paths between groups of concepts in CLO_{ij}. Each group of concepts is treated as a single concept.

(5) *Analyze identified paths:* consists in applying classical Information Retrieval metrics (e.g. recall, precision and f-measure [29]) to determine the level of coverage and conciseness of each path, respectively.

(6) *Determine the ontology summary:* selects the best candidate path according to: (i) f-measure - the path should be the one having the maximum number of relevant concepts and the minimum number of non-relevant concepts; (ii) average relevance - if two distinct paths with the same value of f-measure are identified, the path with the highest average relevance should be preferred.

This work considers that a summary does not represent a cluster ontology in its entirety. Thus, information loss can occur during the summarization process. An evaluation of *OWLSum* is available in [21], where summaries are analyzed in terms of quality. The paper shows how much the result of peer clustering is affected when each cluster of peers is represented by its corresponding ontology summary, instead of its entire cluster ontology. In the following sections, we describe our semantic-based approach for clustering similar peers and for reformulating queries within a community of SPEED.

3 Semantic-Based Peer Clustering

In this section, we describe a semantic-based process for clustering peers in SPEED. Although the proposed process aims at clustering peers in a PDMS, it can be applied to a data integration system or any other distributed system in which data sources communicate via some network protocol. First, an overview of the clustering process is given. Then, a demonstration of how a requesting peer searches for a corresponding community in the structured network is presented. Also, the algorithm for inserting a requesting peer into a semantically similar cluster is described. The steps to connect a requesting peer to an existing cluster and to create a new cluster are also detailed.

3.1   An Overview of the Clustering Process

In SPEED, the connection of a requesting peer is performed in a twofold way (Figure 6). First, a corresponding community is searched in the structured network. If the community is found, then a semantically similar cluster is searched in the unstructured network of the identified community. In both cases, the requesting peer's local ontology is of great importance since it is used to associate the requesting peer to an appropriate community as well as to a semantically similar cluster.

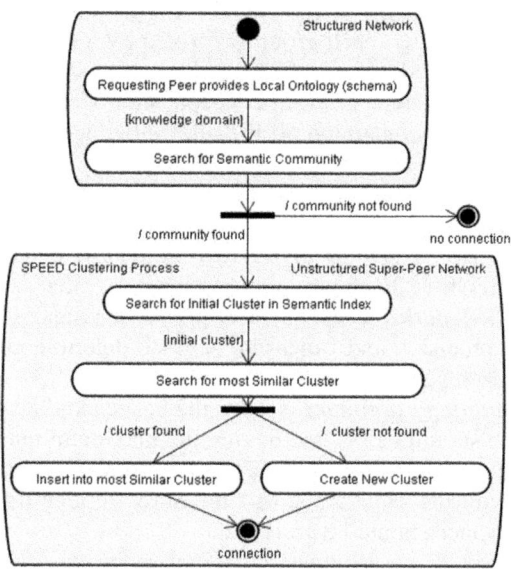

Fig. 6. The several steps involved in the connection of a requesting peer

Assuming that a community has been found by a requesting peer, the search for a semantically similar cluster begins when an initial cluster is provided to the requesting peer. Such initial cluster is obtained from the semantic index of the identified community. The semantic index is located at the semantic peer and corresponds to a data structure summarizing the content available in the clusters of that community. The search for a semantically similar cluster starts at the initial cluster and continues by visiting the semantic neighbors of the initial cluster in the unstructured network. Two distinct clusters, represented by their corresponding integration peer, are considered **semantic neighbors** (neighbors, in short) if they participate in the same community and share similar cluster ontologies.

At each visited cluster, *SemMatcher* is used to determine the global similarity measure between the cluster and the requesting peer. Two peers are **semantically similar** if the global similarity measure between their ontologies is above a certain threshold, called **cluster threshold** (denoted *ct*). The integration peer of each visited cluster returns a global similarity measure to the requesting peer RP_n. RP_n joins the cluster that returns the highest similarity measure (above *ct*) or creates a new cluster. In both cases RP_n connects to the community. Once a requesting peer RP_n is connected, it can assume different roles in the system. For instance, if RP_n joins an existing cluster, it is connected as a data peer. Otherwise, if RP_n creates a new cluster, then it is connected as an integration peer. Next, we detail each step of the proposed semantic-based process for clustering peers in SPEED.

3.2 Search for a Semantic Community

SPEED's structured network is composed of multiple semantic peers and can be built according to any Distributed Hash Table (DHT) protocol, e.g., Chord [17]. A semantic community is described by a set of keywords associated to its corresponding knowledge domain. For instance, the set of keywords related to the *Education* domain can comprise *education* and *university*. Such keywords are defined by a system administrator and are used to locate the *education* community in the DHT network. In order to search for a community, a requesting peer RP_n must first provide an *interest theme* (i.e., an abstract description of its knowledge domain) to an arbitrary semantic peer in the DHT network. An interest theme corresponds to a keyword such as *Education* and *Bioinformatics*. In this current version the interest theme is manually provided by the user (requesting peer). Techniques to automatically determine the interest theme include: (i) provide the concept that is the root of the requesting peer's ontology; (ii) try to infer the domain of the requesting peer's ontology according to its current concepts.

The search for a semantic peer is done incrementally: at each step the successor of a semantic peer is identified until the closest semantic peer is found. Thus, if the provided interest theme is contained in the set of keywords of a community then the corresponding semantic peer will necessarily be found. In this case, a message containing the corresponding semantic peer's address is sent back to the requesting peer. If the interest theme is not found, it probably means that the referred community does not exist in the DHT network or the provided interest theme is not used to describe any community. In this case, the requesting peer should try another interest theme. In this current version, a peer is able to participate in only one community and a keyword cannot be used to describe more than one community.

3.3 Clustering Algorithm

Because of the dynamicity of P2P networks, in SPEED, clustering is mainly an incremental process. Peers are added to clusters one at a time depending on some criterion, e.g., semantic similarity between a requesting peer and current clusters. The clustering process is based on the Leader algorithm [30], which presents some drawbacks when applied to P2P environments such as SPEED. The drawbacks are listed in Table 2 as well as the adaptations made in SPEED.

Table 2. Drawbacks of the Leader algorithm and the corresponding adaptations

Drawbacks of the Leader algorithm	Introduced adaptations
Assume that a centralized view of the clusters is available.	The clusters of each community are connected in an unstructured P2P network and should be searched accordingly. Thus, flooding is used to find clusters in a community.
Clusters are isolated, i.e. there are no links between them.	The clusters are connected through semantic correspondences which are needed to enable query answering. In this sense, we consider the notion of *semantic neighbors* to link clusters only with the most similar ones.
Since clusters are searched in a fixed order, the initial clusters tend to concentrate a high number of peers.	An initial cluster is indicated at each time a requesting peer is to be inserted. The initial cluster is obtained from a semantic index.
The comparison with all clusters may cause scalability and/or performance problems.	We limit the number of clusters to be searched in the unstructured network. Besides the initial cluster, the other clusters to be searched include the *neighbors* of the initial cluster.

The introduction of the proposed adaptations results in a new clustering algorithm (named Leader++) whose pseudo-code is described in Figure 7. Similarly to the Leader algorithm, the proposed algorithm is also order-dependent [31], i.e., it may generate different clusters for different orders of the same input data. It is assumed that a requesting peer RP_n has already discovered a semantic community. In lines 1-2, clustering parameters are initialized, e.g. cluster threshold. The first requesting peer RP_1 is assigned to the cluster CL_1 (line 3). The following requesting peers must search for the most semantically similar cluster. They are addressed to an initial cluster (line 5) and start visiting (some of) the semantic neighbors of the initial cluster (lines 6-8). Each cluster determines the global similarity measure between itself and the requesting peer. RP_n joins the cluster that returns the highest similarity measure that is above *cluster threshold* (lines 9-10) or creates a new cluster (line 11). In both cases, it is necessary to determine the neighbors of the cluster (line 12).

```
01   Let ct (cluster threshold) be a similarity threshold
02   Let connectTTL be a search boundary
03   Let the first requesting peer RP₁ be assigned to cluster CL₁
04   For each requesting peer RPₙ₊₁
05       Search for initial cluster in Semantic Index
06       Start at the initial cluster and while connectTTL > 0 do
07           simClust ← Search for most similar semantic cluster
08           connectTTL ← connectTTL - 1
09       maxSim ← GetMaximumSimilarity(simClust)
10       If maxSim ≥ ct, connect RPₙ₊₁ to the corresponding cluster CLⱼ
11       Else, connect RPₙ₊₁ to a new cluster CLₖ
12       Determine the semantic neighbors of CLⱼ (or CLₖ)
```

Fig. 7. The clustering algorithm Leader++

3.3.1 Search for Initial Cluster in Semantic Index

In SPEED, the connection of requesting peers is continuous and unlimited. Matching a requesting peer's local ontology LO_n against all cluster ontologies CLO_{ij} of a semantic community CM_i is a costly and time-consuming task and therefore should be avoided. The main reasons for that are: (i) the size of cluster ontologies can be large since they integrate multiple local ontologies; and (ii) the number of clusters varies and cannot be predicted.

In order to provide an initial cluster to RP_n, we use a semantic index located at each semantic peer SP_i. In this semantic index, each cluster CL_{ij} of community CM_i is represented by its summarized cluster ontology OS_{ij}. When a requesting peer RP_n finds a community CM_i, its local ontology LO_n is sent to the corresponding semantic peer SP_i.

The search in the semantic index is done by matching LO_n against the summarized cluster ontologies OS_{ij}. For each index entry a global similarity measure between OS_{ij} and LO_n is produced by *SemMatcher*. Afterwards, SP_i determines the initial cluster by ranking the computed global similarity measures in descending order. The initial cluster will be the one associated with the highest global measure. Finally, the corresponding integration peer's address is returned to RP_n. If no initial cluster is identified (for example, if the community is empty) then RP_n creates a new cluster. In this case, RP_n connects as an integration peer.

3.3.2 Search for the Most Semantically Similar Cluster

Given an initial cluster, the problem now is to determine the clusters in the community CM_i that should be visited in order to search for a semantically similar cluster. To this end, the semantics of the involved clusters is taken into account through the notion of semantic neighbors. A cluster CL_{ij} is a semantic neighbor of CL_{ik}, if the global similarity measure between CLO_{ij} and CLO_{ik} is above a certain threshold called **neighbor threshold** (denoted nt). Given a cluster CL_{ij} and its semantic neighborhood N_{ij}, a semantic cluster $CL_{ik} \in N_{ij}$ is such that $SemMatcher(CLO_{ij}, CLO_{ik}) \geq nt$.

In addition, given the neighborhood $N_{ij} = \{CL_{i1}, CL_{i2},...,CL_{ik}\}$ of a cluster CL_{ij}, all the clusters in N_{ij} are considered **direct neighbors** of CL_{ij}. If a cluster CL_{in} is not included in N_{ij} but is contained in N_{ik} (i.e., the neighborhood of CL_{ik}) then we say that CL_{in} is an **indirect neighbor** of CL_{ij}. In Figure 8, CL_{i3} and CL_{i4} are direct neighbors of CL_{i2} and indirect neighbors of CL_{i1}.

Based on the notion of semantic neighbors, several possible search strategies can be derived in order to limit the number of clusters to be searched in a community. All of them can be controlled by a TTL limit (denoted *connect TTL*). For instance, if:

- *connect TTL* = 1, the search scope is limited to the initial cluster;
- *connect TTL* = 2, the search scope includes the initial cluster and its direct semantic neighbor(s);
- *connect TTL* $\geq$ 3, the search scope includes the initial cluster as well as its direct and indirect semantic neighbor(s).

The search starts when RP_n sends its local ontology LO_n to the integration peer corresponding to the initial cluster. At the integration peer, *SemMatcher* is executed by taking as arguments the current cluster ontology and LO_n. The resulting global similarity measure is returned to RP_n. According to the defined search strategy, LO_n can be propagated to the direct and/or indirect semantic neighbors of the initial cluster. At each visited cluster, *connect TTL* is decreased and the search process continues. The search finishes when *connect TTL* reaches zero.

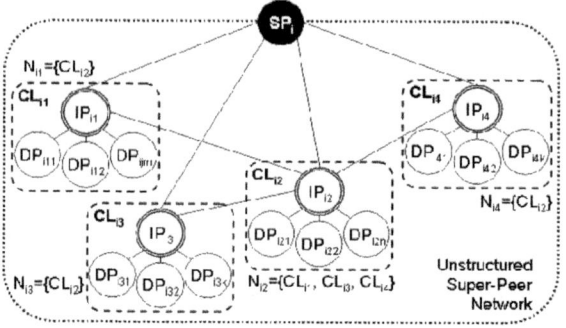

Fig. 8. An instantiation of a semantic community

Figure 9 is a sequence diagram illustrating the search for the most similar cluster in the community of Figure 8. Since *connect TTL* is set to 3, the four clusters (CL_{i1}, CL_{i2}, CL_{i3}, and CL_{i4}) are visited in order to determine the most similar cluster to the requesting peer RP_n. The clusters CL_{i1}, CL_{i2}, CL_{i3}, and CL_{i4} are represented by their corresponding integration peers IP_{i1}, IP_{i2}, IP_{i3}, and IP_{i4}, respectively. The first visited integration peer is IP_{i1} that corresponds to the initial cluster provided by the semantic peer SP_i. The search scope comprises the direct (IP_{i2}) and indirect (IP_{i3} and IP_{i4}) semantic neighbors of IP_{i1}. The global similarity measures returned to RP_n are: 0.5 (IP_1), 0.6 (IP_2), 0.2 (IP_3), and 0.3 (IP_4). These measures are used by RP_n to decide whether to join one of the visited clusters or to create a new one (in this case, CL_{i5}). Once RP_n receives the global similarity measures from the visited clusters, RP_n must select the highest global similarity measure. Thus, two possibilities can occur:

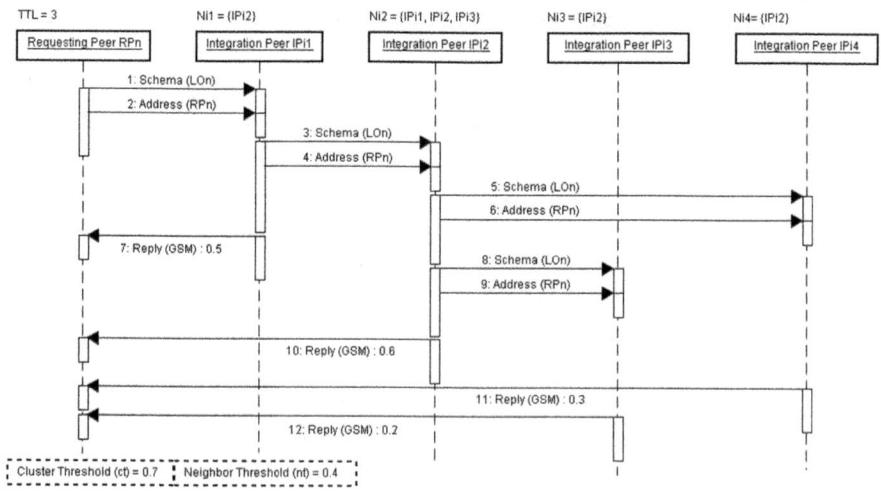

Fig. 9. A requesting peer RP_n searches for a semantically similar cluster

Case 1: If the selected measure is equal or higher than *cluster threshold* (*ct*), then RP_n joins the corresponding cluster as a data peer. In this case, **OntMerger**, the ontology merging tool of SPEED, is used to merge the current CLO_{ij} and the local ontology LO_n of the new data peer. As a result, a new version of CLO_{ij} is produced as well as a set of semantic correspondences between CLO_{ij} and LO_n which are needed for query answering. In addition, a new summary of CLO_{ij} (OS_{ij}) is built by **OWLSum** and the semantic index is updated accordingly.

Case 2: If the selected measure is lower than the *cluster threshold* (*ct*), RP_n creates a new cluster and joins that cluster as an integration peer. In this case, the cluster ontology CLO_{ij} of the new cluster corresponds to the local ontology LO_n describing RP_n. The semantic neighborhood of the new cluster is composed of all the visited clusters CL_{ik} such that $nt \leq SemMatcher(CLO_{ij}, CLO_{ik}) < ct$. A summarized version of CLO_{ij} (OS_{ij}) is built by **OWLSum** and a new entry is added to the semantic index. In Figure 9, *neighbor threshold* (*nt*) and *cluster threshold* (*ct*) are set to 0.4 and 0.7, respectively. Since the highest global similarity measure returned by the searched clusters (i.e., 0.6, returned by CL_{i2}) is lower than *ct*, the requesting peer RP_n will create a new cluster (CL_{i5}). The semantic neighborhood of CL_{i5} is defined as $N_{i5} = \{CL_{i1}, CL_{i2}\}$.

4   Semantic-Based Query Reformulation

One key issue regarding query answering in SPEED is the reformulation of a query posed at a peer into another one over a target peer. Two aspects should be considered when dealing with query reformulation. First, querying distributed data sources should be useful for users, i.e., resulting query answers should be in conformance with users' preferences. A second aspect is that concepts from a source peer do not always have exact corresponding concepts in a target one. This may result in an empty reformulation and, possibly, no answer to the user. Regarding the former aspect, we argue that user preferences and the current status of the environment should be taken into account at query reformulation time; regarding the latter, the original query should be adapted to bridge the gap between the two sets of concepts.

In SPEED, we propose a query reformulation approach – named *SemRef*, which uses semantics as a way to better deal with the aforementioned aspects. In order to capture user preferences, query semantics and environment parameters we use *context* [11, 14]. We accomplish query reformulation and adaptation through *query enrichment*, i.e., we analyze an initial query expression (posed in a source peer) in order to find out some extra semantic knowledge that can be added (in a target peer) at query reformulation time, so its resolution may provide expanded (additional related) answers. To this end, besides equivalence, we use other correspondences which go beyond the ones commonly found (specialization, generalization, aggregation), proposing disjointness and closeness, identified by *SemMatcher* (see Figure 5). Through this set of semantic correspondences, we produce two kinds of query reformulations: (i) an *exact* one, considering equivalence correspondences; and (ii) an *enriched* one, resulting from the set of the other correspondences. The priority is to produce the best query reformulation through equivalence correspondences, but if that is not possible, or if users define that it is relevant for them to receive semantically related answers, an enriched reformulation is also generated. As a result, users are provided with a set of expanded answers, according to their preferences.

In this work, to simplify matters, we address query reformulation in a setting based on two peers, although in a P2P environment the reformulation process can be applied to pairs of peers composing a path of the query routing process. In our work, we are not concerned with view-based query rewriting as works which deal with GAV/LAV strategies to reformulate queries posed through a global schema [1, 9]. Our problem is different of the query rewriting problem because we focus on reformulating a query posed at a source schema in terms of a target schema by replacing the original concepts of the query by semantically related concepts from the target schema according to some acquired contextual information such as user preferences.

4.1 Using Semantics in Query Reformulation

In a dynamic environment such as SPEED, semantics may be identified considering the user's and/or peers' perspectives or even the query formulation. In our setting, the peers are grouped within the same knowledge domain (e.g., *Education*), which enables us to use domain ontologies (DO), as background knowledge to identify correspondences between the peer ontologies [8]. These peer ontologies are terminologically normalized according to the DO. As a result of a semantic matching process, as explained in Section 2.4, we have a set of semantic correspondences between pairs of neighbor peers which are used at query reformulation time. To clarify matters, in Section 4.3, we provide an example which presents two cluster ontologies and a set of identified semantic correspondences between them.

Another kind of semantic knowledge we use is context [11, 13]. When a user poses a query, all the surrounding contextual elements will be analyzed in order to select the data that is relevant to the user's specific situation. In our approach, we use three types of context: of the user, of the query, and of the environment. In SPEED, for each submitted query, the whole query execution process instantiation changes according to the context of the query, of the peers, of the semantic correspondences and of the user [12]. This contextual information is stored in a contextual ontology, named CODI [12] which allows the system to organize knowledge (i.e., contextual elements), recognize conditions and assist query answering.

The context of the users is acquired when they state their preferences concerning the reformulation strategy. These preferences involve setting four variables which specify what should be considered when Q is to be enriched. The variables are defined as follows:

- *Approximate:* includes concepts that are close to the ones of Q;
- *Specialize:* includes concepts that are sub-concepts of some concepts of Q;
- *Generalize:* includes concepts that are super-concepts of some concepts of Q;
- *Compose:* includes concepts that are part-of or whole-of some concepts of Q.

The context of the query is obtained through the analysis of its semantics (e.g., operator to be applied) and through the query reformulation mode (defined by the user). In the former, the query concepts and constructors are identified. The latter relates to the way the reformulation algorithm operates: *expanded*, where both exact and enriched reformulations may be provided, or *restricted*, where the priority is to produce an exact reformulation, an enriched reformulation is provided only if the exact reformulation gives no result.

The context of the environment is acquired at two different times: (i) when the user sets the variable *Path_Length*, which limits the number of subsequent reformulations in the set of integration peers; and (ii) when the system identifies in which integration peer the query has been submitted and also establishes the context of the neighbor peers to where the queries will be reformulated and routed.

Most contextual information used in this work is acquired at query submission time. Some are gathered from the users' preferences, while others are inferred on the fly according to the environment's conditions. However, in this paper, we only deal with the context acquired from the user preferences and from the query itself.

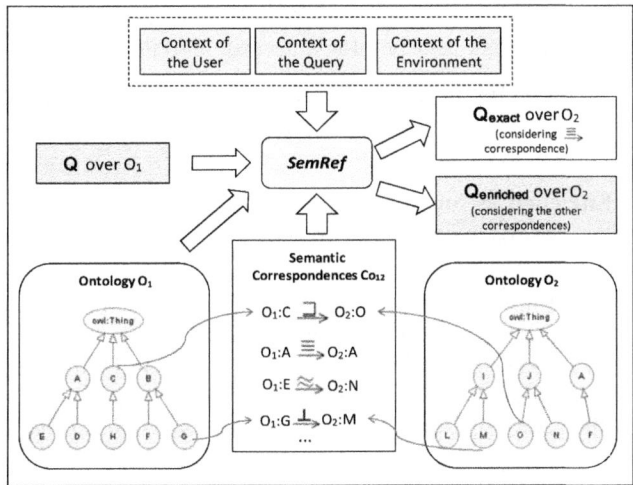

Fig. 10. The SemRef approach

Considering these semantic elements, the principle of our approach is to enhance query reformulation by using them in such a way that we can provide users with a set of expanded answers. As depicted in Figure 10, when a query Q is submitted to an integration peer P_1, *SemRef* considers the semantic correspondences (Co_{12}) between the source and target ontologies (O_1 and O_2) along with the concerned contextual information and produces two types of reformulations: Q_{exact} and $Q_{enriched}$. Exact and enriched query reformulations are produced as a means to obtain expanded answers.

4.2 The *SemRef* Algorithm

Our query reformulation approach has been encoded in $\mathcal{ALC}$-$\mathcal{DL}$ [32]. Thus, a query Q over a peer ontology O_1 is a concept expression, $Q = C$, where C is an $\mathcal{ALC}$ concept. An $\mathcal{ALC}$ concept may be an atomic concept or a complex concept including roles, quantifiers, conjunctions or disjunctions. In our work, we consider that a query Q is a formula consisting of a disjunction of queries which are themselves conjunctions of $\mathcal{ALC}$ concepts $C_1, \ldots, C_n$ where $n \geq 1$.

Definition 2 - Query. A query Q expressed over a peer ontology, has the following form: $Q = Q_1 \sqcup Q_2 \sqcup ... \sqcup Q_m$, where $Q_i = C_1 \sqcap C_2 \sqcap ... \sqcap C_n$, and where each C_j is an atomic concept, a negated atomic concept of a quantified atomic concept (C_j, $\neg C_j$, $\forall R.C_j$ or $\exists R.C_j$).

A query example following such definition is Q = [Teacher $\sqcap$ Researcher] $\sqcup$ [Student $\sqcap$ Researcher] which asks for people who are *teachers* and *researchers* or *students* that are also *researchers*.

We argue that, when posing a query, users must be aware that not only exact answers, but also those that meet or complement their initial intention, can be relevant for them. The algorithm verifies users' preferences. If these preferences enable query enrichment, besides the exact reformulation, it produces an enriched one, providing users with answers at different levels of closeness. In this light, a query formulated in terms of a source peer ontology may be reformulated exactly or approximately into a query using terms of a target peer ontology, according to the following definitions:

Definition 3 - Exact Reformulation. A reformulation Q' of a query Q is said to be exact (denoted as Q_{exact}) if each concept (or property) C' of Q' is related to a concept (or property) C of Q by a Co correspondence, where Co $\in$ { $\overset{\equiv}{\rightarrow}$ } (equivalence).

Definition 4 - Enriched Reformulation. A reformulation Q' of a query Q is said to be enriched ($Q_{enriched}$) if each concept (or property) C' of Q' is related to a concept (or property) C of Q by a Co correspondence, where Co $\in$ { $\overset{\sqsubseteq}{\rightarrow}$, $\overset{\sqsupseteq}{\rightarrow}$, $\overset{\approx}{\rightarrow}$, $\overset{\rhd}{\rightarrow}$, $\overset{\lhd}{\rightarrow}$, $\overset{\perp}{\rightarrow}$ }.

The *SemRef* algorithm tries to produce the set of exact reformulations, although sometimes the produced set is empty. Enriched reformulations will be produced in two situations: (i) if the user requires them through the set of enriching variables (*approximate*, *specialize*, *generalize*, and *compose*) and *expanded* query reformulation mode; or (ii) if the reformulation mode was defined as *restricted*, but the exact reformulation gave no result.

The *SemRef* algorithm receives as input a query Q (at integration peer P_1), the target integration peer P_2, the set of semantic correspondences between them and the acquired context (e.g., enriching variables and mode values). As output, it produces one or two reformulated queries (Q_{exact} and/or $Q_{enriched}$). The *SemRef* algorithm is sketched in Figure 11. In order to obtain the reformulations, the algorithm performs the following tasks:

I.   It receives query Q (a disjunction of conjunctions of ALC concepts). For each conjunctive query Q_k in Q, while there are concepts C_j in Q_k to process, it adds the corresponding concepts to one of three sets:

- S_1C_j: the set of concepts that are equivalent to C_j.
- S_2C_j: the set of concepts related to C_j by other kinds of correspondences (closeness, specialization, generalization, part-of, and whole-of).
- Neg_S_2C_j: if there is a negation over C_j, *SemRef* searches for disjointness correspondences to directly get the opposite concept. In this case, the concept is added to Neg_S_2C_j set. If there is no disjointness correspondence, a variable BNeg is set to TRUE and later in the algorithm, the negation is done over the corresponding concept found through the set of

other semantic correspondences (equivalence, specialization, generalization, part-of, whole-of or closeness).

II. After processing all the concepts of a conjunctive query, *SemRef* verifies if there were exact correspondences and if the conjunction did not fail (i.e., all existing concepts in the conjunction had corresponding ones). If so, it builds the exact reformulation for the current conjunctive query Q_k (line 33).

SemRef (Q, P₁, P₂, Co[P₁,P₂], MODE, REF_VAR, Qexact, Qenriched)
1.   For each Qk in Q /* for each conjunctive query in Q */
2.   B ← TRUE /* used to stop the search if some concept has no correspondent one in P₂ */
3.   While (there is still a concept Cj in Qk to process) and (B=TRUE)
4.   S1Cj ← ∅ /* set of concepts that are equivalent or subconcept of Cj */
5.   S2Cj ← ∅ /* set of concepts related to Cj by other correspondence except disjointness*/
6.   Neg_S2cj ← ∅ /* set of concepts related to Cj by disjointness correspondence */
7.   For each isEquivalentTo or isSubConceptOf assertion between Cj and a concept C'
8.   Add C' to S1Cj
9.   End For; /* End of the loop related to the assertions equal to $\stackrel{\equiv}{\rightarrow}$ or $\stackrel{\sqsubseteq}{\rightarrow}$ */
10.   For each other kind of assertion involving Cj
11.   If (MODE is expanded) or (MODE is restricted and S1Cj is empty) Then
12.   If APPROXIMATE = TRUE Then
13.   If there is a concept C' in P' such that C' $\stackrel{\approx}{\rightarrow}$Cj Then
14.   Add C' to S2Cj
15.   If GENERALIZE = TRUE Then
16.   If there is a concept C' in P' such that C' $\stackrel{\sqsupseteq}{\rightarrow}$Cj Then
17.   Add C' to S2Cj
18.   If COMPOSE = TRUE Then
19.   If there is a concept C' in P' such that C' $\stackrel{\triangleright}{\rightarrow}$Cj or C' $\stackrel{\triangleleft}{\rightarrow}$ Cj Then
20.   Add C' to S2Cj
21.   If Cj is negated Then
22.   If there is a concept C' in P' such that C' $\stackrel{\perp}{\rightarrow}$Cj Then
23.   Add C' to Neg_S2Cj
24.   BNeg ← TRUE
25.   End For; /* End of the loop related to the assertions different from $\stackrel{\equiv}{\rightarrow}$ or $\stackrel{\sqsubseteq}{\rightarrow}$*/
26.   If (S1Cj = ∅ and S2Cj = ∅ and Neg_S2C2 = ∅) Then B ← FALSE
27.   End While; /* End of the loop processing concepts */
28.   B1 ← TRUE;
29.   If (one of S1Cj = ∅) Then B1 ← FALSE /* the conjunction fails */
30.   B2 ← TRUE;
31.   If (one of S2Cj = ∅) Then B2 ← FALSE /* the conjunction fails */
32.   If B1 = TRUE Then
33.   Qk_exact ← Build_Exact_Reformulation (Qk, S1C1, S1C2, ..., S1Cp)
34.   Else Qk_exact ← ∅
35.   If B2 = TRUE or BNeg = TRUE Then
36.   Qk_enriched ← Build_Enriched_Reformulation(Qk, S2C1, ... S2Cp, Neg_S2C1, ... Neg_S2Cp)
37.   Else Qk_enriched ← ∅
38. End For; /* End of the loop processing the conjunctive queries Qk */
39. If (at least one of Qk_exact ≠ ∅) /* at least one of Qk's exact reformulations is not empty */
40.   Then Qexact ← Build_Final_Exact_Reformulation (Q, Q1_exact, ..., Qm_exact)
41.   Else Qexact ← ∅
42. If ((MODE is expanded) or (MODE is restricted and Qexact is empty)) and
43.   (at least one of Qk_enriched ≠ ∅)
44.   Then Qenriched ← Build_Final_Enriched_Reformulation (Q, Q1_enriched,..., Qm_ enriched)
45.   Else Qenriched ← ∅
46. End
End_SemRef;

Fig. 11. A high level view of the *SemRef* algorithm

III.   If there were enriching correspondences and the conjunction did not fail, then *SemRef* builds the enriched reformulation for the current conjunctive query Q_k (line 36).

IV.   Finally, after processing all the conjunctive queries Q_k of Q, SemRef produces the final Q_{exact} (line 40), as the disjunction of the resulting exact conjunctions and the final $Q_{enriched}$ as the disjunction of the resulting enriched conjunctions (line 44).

The main properties of the *SemRef* algorithm are: (i) soundness, meaning that every produced query reformulation is a "correct" reformulation solution; and (ii) completeness, implying that it always gives a solution when there is one. For the formal proofs of soundness and completeness, we refer the reader to [33].

4.3 *SemRef* in Practice

Our example scenario is the one presented in Section 2.3. Since terminological normalization is a pre-matching step in which the initial representation of two ontologies are transformed into a common format suitable for similarity computation, we have normalized both ontologies O_1 and O_2 to a uniform representation format according to the DO. In order to identify the semantic correspondences between O_1 and O_2, the rules of *SemMatcher* were applied [20]. As a result, the set of semantic correspondences between O_1 and O_2 were identified. Since the correspondences are unidirectional, we present examples of this set concerning the concept **FullProfessor** (from O_1) with some related concepts in O_2:

- O_1:FullProfessor $\xrightarrow{\equiv}O_2$:FullProfessor (isEquivalentTo)
- O_1:FullProfessor $\xrightarrow{\sqsubseteq}O_2$:Professor (isSubConceptOf)
- O_1:FullProfessor $\xrightarrow{\approx} O_2$:VisitingProfessor (isCloseTo)
- O_1:FullProfessor $\xrightarrow{\perp}O_2$:AssociateProfessor (isDisjointWith)
- O_1:FullProfessor $\xrightarrow{\triangleright} O_2$:Course (isPartOf).

In this illustrative set, we can see the equivalence correspondence between **FullProfessor** in O_1 and O_2. This is the most commonly identified correspondence type in traditional query reformulation approaches. On the other hand, by using the semantics underlying the DO, we can identify the other unusual correspondences (e.g., closeness).

We present the *SemRef* main steps in practice through the query Q = **FullProfessor**, submitted in P_1. *SemRef* starts by initializing the sets S_1C_1, S_2C_1 and Neg_S_2C_1. The first set receives the concepts that are equivalent to FullProfessor, i.e., S_1C_1 = {FullProfessor}. The second set receives the concepts resulting from the other correspondences (except disjointness), i.e., S_2C_1 = {VisitingProfessor, Professor, Course}, considering that the user has set all four enriching variables to TRUE and the reformulation mode to EXPANDED. The third set would receive disjoint concepts, if there was a negation over the concept FullProfessor. Since the query is composed of only one concept and there is no negation over it, the algorithm verifies that both sets (S_1C_1 and S_2C_1) are not empty and consequently builds both exact and enriched reformulations. The final exact reformulation is Q_{exact} = [FullProfessor]. The enriched one is $Q_{enriched}$ = [VisitingProfessor ⊔ Professor ⊔ Course].

Another example regards the submitted query Q = ¬UndergraduateStudent (in P_1). Suppose that the user has set the *specialize* variable to TRUE, and the *restricted* option for the reformulation mode. Also, assume that there is no equivalent concept of UndergraduateStudent in P_2, which entails S_1C_1 = { }. Since query reformulation mode was defined as restricted, but S_1C_1 was empty, enrichment is considered by *SemRef*. Thus, S_2C_1 set receives {Monitor}, according to the specialization correspondence. Because there is a negation over the concept UndergraduateStudent, the third set Neg_S_2C_1 is set to {Worker, GraduateStudent}, according to disjointness correspondences. Thereby, although the algorithm does not build an exact reformulation, it produces an enriched one, negating over the concept *Monitor*, and providing a union of such negation with the concepts *Worker* and *GraduateStudent* (from Neg_S_2C_1 set). The final exact reformulation is Q_{exact} = { }, but the final enriched one is $Q_{enriched}$ = [¬Monitor ⊔ Worker ⊔ GraduateStudent].

5 Experiments and Results

In this section, we discuss implementation issues and describe conducted experiments regarding our approach for peer clustering and query reformulation.

5.1 Peer Clustering

For our experiments, we have developed a PDMS simulator[3] through which we were able to reproduce the main conditions characterizing the SPEED's environment. The simulator was implemented in Java. Through the simulator we were able to generate scenarios corresponding to networks of peers, each with its own schema describing a particular reality. We assume that there exists a communication link among the peers that enables sending and receiving information, i.e. queries, data, and schema information. In this current version, the tool is able to simulate requesting peer connection and the formation of clusters in a given semantic community. In order to execute the experiments, we have included in the simulator the ontology management tools *SemMatcher*, *OntMerger*, and *OWLSum*.

Our simulation tests were conducted considering the *education* knowledge domain. Therefore, we have manually built an ontology library containing small local ontologies to be used by requesting peers during the tests. A description of all ontologies is available in the simulator's website (http://www.cin.ufpe.br/~speed/simulator/index.htm). The local ontologies were derived from the real-world ontology *UnivCSCMO.owl* (Section 2.3). During our tests, we have assumed that the element names of the local ontologies were normalized according to the element names of the community ontology.

All tests were performed in an Intel Pentium M 1.60GHz, 1GB of RAM. The operating system was Windows XP®. In our experiments, the SPEED's DHT network was first created with ten semantic peers. Afterwards, we started the connection of

---

[3] http://www.cin.ufpe.br/~speed/simulator/index.htm

requesting peers one at a time. Each requesting peer has searched for a corresponding semantic community (*education*) in the DHT network and then for a semantically similar cluster in the unstructured network of the discovered community.

We have evaluated the clustering results using classical external statistical indices: *Rand Index* [34], *Jaccard Coefficient* [35], *Fowlkes-Mallows (FM) Index* [36], and *Hubert's statistic* [37]. The indices were computed through a comparison between the clustering results obtained from the simulator against an ideal one generated by a hierarchical clustering algorithm [38]. The hierarchical algorithm follows the *batch approach* for clustering a set of peers. In other words, it considers the set as a whole and begins to organize peers into meaningful clusters. A clustering result refers to the network generated after a successful execution of a simulation test.

The SPEED's clustering algorithm, Leader++, is incremental and order-dependent, as explained in Section 3.2. Due to the second characteristic, the indices were calculated multiple times considering different orders of requesting peers. Then, for each of the statistical indices an average of the index results was calculated. The values of the statistical indices lie between 0 and 1. However, a requirement for achieving the maximum value is to have the same number of clusters in both clustering results, which, as observed, is not always possible. For all the used indices, the larger their value the higher the agreement between the two clustering results.

The goal of our experiments is to verify whether semantically similar peers are being forwarded to appropriate clusters. Two types of experiments were performed. In the first one, we considered the search strategy proposed in SPEED (denoted here *limitedClusters*). According to such strategy, a requesting peer receives an initial cluster and visits only a limited number of clusters, i.e., the neighborhood of the initial cluster. In the second experiment (denoted *allClusters*), we considered a different search strategy. Each requesting peer visits all current clusters before connecting to the system. In this case, the semantic index is discarded.

For the *limitedClusters* strategy, the following parameters (with respective values) were considered: *summary size* (6 concepts), *cluster threshold* (0.35), *neighbor threshold* (0.10), and *connectTTL* (3 hops). For the *allClusters* strategy, the clustering result was also compared with the ideal one. These values are an estimative and were defined based on the observation of the clustering algorithm behavior considering the set of ontologies used in the experiments. To guarantee that all clusters were visited we have modified the parameters *connectTTL* (999) and *neighbor threshold* (0). *Summary size* and *cluster threshold* remained unaltered. Afterwards, we compared the obtained index results against the best ones of *limitedClusters*. Naturally, *allClusters* tends to produce better index results than *limitedClusters*. However, we have obtained similar index results (Figure 12) with fewer executions of *SemMatcher* (Figure 13).

The decrease is explained because when a requesting peer arrives at a semantic community, only a limited number of clusters are visited in order to determine the most similar cluster for the requesting peer. Consequently, the number of executions of *SemMatcher* is minimized. The numbers available in Figure 13 indicate a reduction of 27% of matchings involving local ontologies and cluster ontologies, and a reduction of 25% of matchings between cluster ontologies. This reduction optimizes not only the clustering process but also que query routing process. Since the semantic index was discarded in the *allClusters* strategy, no matchings involving local ontologies and summarized cluster ontologies were computed.

Internal validity was evaluated using the Silhouette indices [39]. Such indices are useful when compact and clearly separated clusters are desired. In this case, there are two interesting issues to be analyzed in a clustering result: the homogeneity of each cluster and the degree of separation between the obtained clusters. The higher their homogeneity and the separation the better is the clustering result. Both aspects can be captured in a global Silhouette value between −1 (bad clustering) and 1 (very good clustering) [39]. Figure 14 illustrates the global Silhouette values for different values of *cluster threshold*. For the set of ontologies used in the experiments, the best clustering result (0.505) for *limitedClusters* has been obtained when *cluster threshold* was set to 0.35.

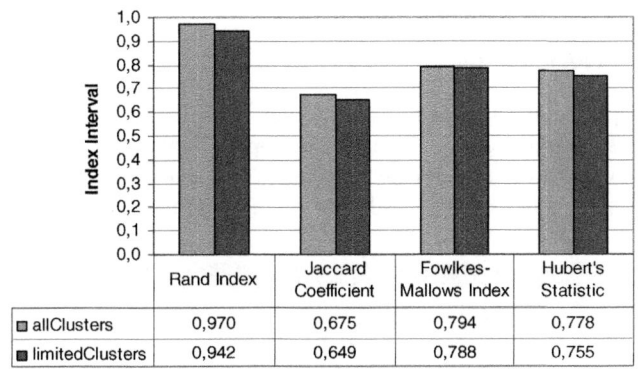

	Rand Index	Jaccard Coefficient	Fowlkes-Mallows Index	Hubert's Statistic
allClusters	0,970	0,675	0,794	0,778
limitedClusters	0,942	0,649	0,788	0,755

Fig. 12. A comparison of search strategies using internal statistical indexes

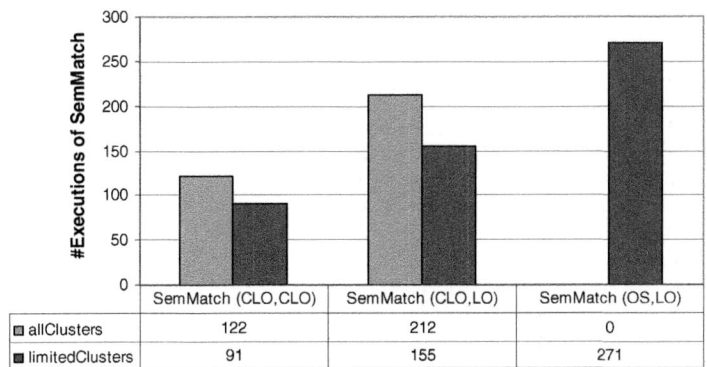

	SemMatch (CLO,CLO)	SemMatch (CLO,LO)	SemMatch (OS,LO)
allClusters	122	212	0
limitedClusters	91	155	271

LO = Local Ontology   CLO = Cluster Ontology   OS = Ontology Summary

Fig. 13. A comparison of search strategies considering the number of executions of *SemMatcher*

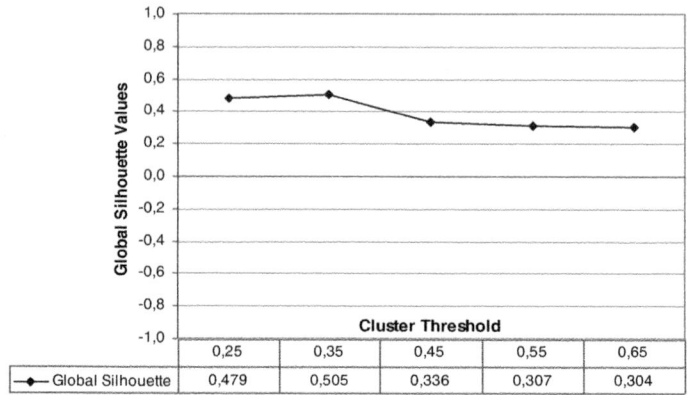

Fig. 14. Clustering validity: internal indices

5.2 Query Reformulation

We have implemented the *SemRef* approach in SPEED using Java. RMI[4] has been used for peer communication. We have adopted both Jena[5] and Protégé's API[6] in order to manipulate the underlying ontologies and execute queries over them. Figure 15 shows a screenshot of the module's main window that is split into three parts:

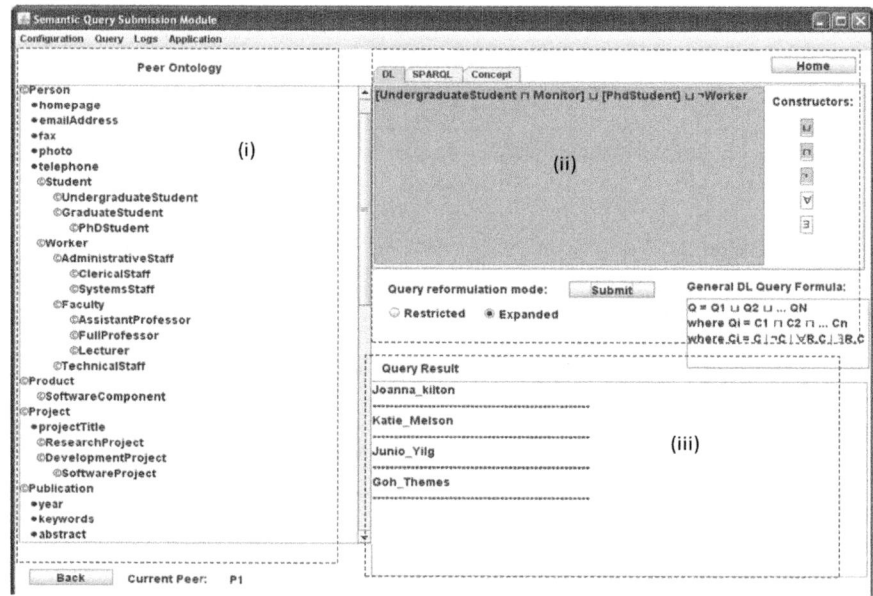

Fig. 15. Query interface with DL Query Formulation option

---

[4] http://java.sun.com/j2se/1.4.2/docs/guide/rmi/
[5] http://jena.sourceforge.net/
[6] http://protege.stanford.edu/

(i) the peer ontology area; (ii) the query formulation area; and (iii) the query results area. Queries can be formulated using SPARQL[7] or using $\mathcal{ALC}$-DL.

In order to perform our experiments, we have considered two integration peers – P_1 and P_2 with ontologies representing the *Education* knowledge domain (discussed in Section 2.3). The set of semantic correspondences between P_1 e P_2 schemas (ontologies) has been generated by SemMatcher and stored in RDF files. We have conducted our evaluation using 55 queries expressed in $\mathcal{ALC}$/DL representing a variety of query possibilities, namely: queries with one concept, queries with negation over a concept, queries with conjunctions, queries with disjunctions, and queries with disjunctions of conjunctions. For each submitted query, we checked the combination of reformulation possibilities according to the set of enriching variables specification and reformulation mode. Examples of such queries are: Q_9: ¬PhDStudent and Q_{30}: [Student ⊓ Monitor] ⊔ [GraduateStudent ⊓ MasterStudent].

The goal of our experiments is to assess the use of semantics in our query reformulation approach. To this end, we have used *precision* (*P*) and *recall* (*R*) [29] as the main metrics. They are usually employed to measure the results of a query execution and to compare them to ideal results expected by the user. In our case, we aim to evaluate the query reformulation phase. Consequently, we are concerned with the quality (correctness) of query reformulation instead of the query execution results (query answers) which currently are outside of our scope. We consider *P* and *R* measures as follows:

$$R(\text{Correct, PossibleRef}) = \frac{\#\,\text{Correct}}{\#\,\text{PossibleRef}} \qquad P(\text{Correct, ProducedRef}) = \frac{\#\,\text{Correct}}{\#\,\text{ProducedRef}}$$

where, given an original query Q, #Correct is the number of correct reformulations (*Exact* or *Enriched*) of Q, #PossibleRef is the total of all possible reformulations of Q and #ProducedRef is the total of all produced reformulations of Q.

We evaluated the produced query reformulations in three ways: (i) *Without* semantics, where *SemRef* reformulates each query without considering any semantics (i.e., enriching variables are disabled and query reformulation mode operates on its default); (ii) *With* semantics, in *restricted* mode, where, at least one enriching variable is set, allowing the algorithm to verify the possibility to produce enriched reformulations in case of empty exact reformulations; and (iii) *With* semantics, in *expanded* mode, where the algorithm tries to produce both exact and/or enriched reformulations.

We have observed that concepts that only exist in one of the peer ontologies usually do not have an equivalent concept in the target one, thus entailing an empty exact reformulation. In these cases, enriching the reformulation has been essential; otherwise, no reformulation query would be obtained. Thereby, as shown in Figure 16, regarding *recall*, we verify that it increases when we apply semantics, i.e., *SemRef* is able to produce a higher number of correct (i.e., *exact* and/or *enriched*) query reformulations. Regarding *precision*, results show that it also increases with the use of semantics. Moreover, we verify that, considering semantics, the *SemRef* algorithm is able to provide the complete set of query reformulations for a given query **Q** [33].

---

[7] http://www.w3.org/TR/rdf-sparql-query/

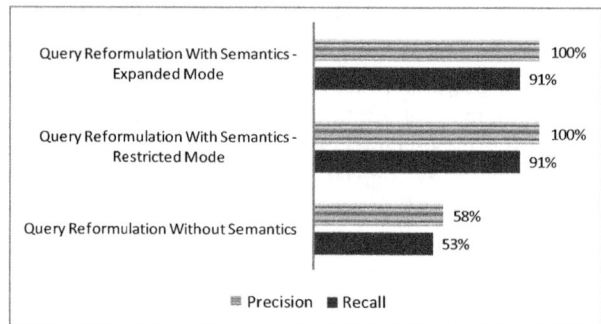

Fig. 16. Degree of precision and recall of query reformulations

Even enabling only one of the enriching variables, the use of the reformulation approach has yielded promising results. Figure 17 presents one query example, considering only *approximate = TRUE*, where *SemRef* produces both exact and enriched reformulations (in this case, only GraduateStudent has an equivalent concept in P₂). Besides, when users set at least one of the enriching variables, they are also defining that the negation over concepts must be dealt, not only with the usual correspondences, but, particularly with disjointness (GraduateStudent *isDisjointWith* UndergraduateStudent). This means that we are able to directly obtain the disjoint concept as a solution to the negation of an original concept. Considering the used peer ontologies, we have assessed in which degree each one of the enriching options (Specialize, Generalize, Approximate, and Compose) assist query enrichment. Precision and recall results considering individually each enriching option are shown in Figure 18.

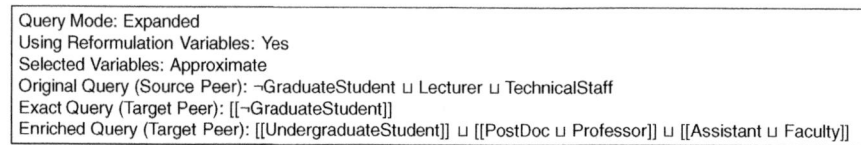

Fig. 17. Exact and Enriched Reformulation with only one Enriching Option

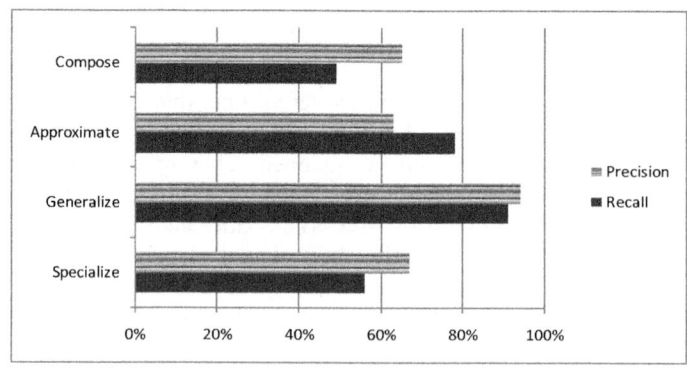

Fig. 18. Degree of precision and recall for each enriching option

Considering a dynamic distributed environment such as ours, there are, at least, three context-dependent factors which influence the answer to a given user query: (i) the current status of the network, mainly in what concerns available peers; (ii) the semantics underlying the submitted query and (iii) the users' preferences regarding the query answering process. As a result, we can see that the context has an impact on the produced answer, and thereby, we cannot guarantee that the complete set of answers will always be returned to users, although we guarantee that the *SemRef* algorithm is complete (i.e., all existing solutions of reformulations are produced), according to the theorem proved in [33]. On the other hand, considering these contextual elements, we can enrich queries, and thus produce a set of answers that match users' preferences while hold environmental conditions.

Although the use of semantics is highly context-dependent, considering our particular experimental setting (the scenarios with their respective peer ontologies and the set of submitted queries), we can conclude that a query reformulation process which is carried out without considering semantics may produce a larger set of empty query reformulations and possibly empty query results for a given query. Using semantics, exact and/or enriched query reformulations may be produced, thus providing a larger set of query reformulation possibilities. We proved that such produced query reformulations are correct. They are sound and complete, according to the acquired context and to the set of semantic correspondences between the current peers. More details on the experiments carried out can be found in [33].

6 Related Work

Some existing PDMS have used semantics for enriching their services, such as Piazza [1, 40], Edutella [41], Xpeer [42], and SenPeer [43]. Furthermore, there has been a lot of recent interest in using ontologies as a way to semantically enrich PDMS services. Helios [7] is a system for ontology-based knowledge discovery and sharing in peer based open distributed systems. The knowledge sharing and evolution processes are based on peer ontologies, which describe the knowledge of each peer, and on interactions among peers. Besides, the system uses a Semantic Matchmaker module in order to identify the semantic affinity among concepts stored in the peer ontologies. Humboldt Discoverer [6] is an extensible PDMS where each peer has a schema against which queries can be posed, and each peer has mappings to one or more peers. Additionally, a peer can have a mapping to one or more ontologies. The system is based on an index concept which enables a peer to locate information sources in a PDMS which are not reachable by schema mappings. SomeRDFS has been developed using a data model based on RDF on top of the SomeWhere infrastructure [44]. Some-Where is a PDMS where there are neither super-peers nor a central server having the global view of the overlay network. Query reformulation is reduced to consequence finding over logical propositional theories solved by DECA (Decentralized Consequence finding Algorithm) – the algorithm of SomeWhere, where each peer possesses a set of propositional clauses built from a set of propositional variables. OntSum [45] is a PDMS which uses an ontology-based approach to address the routing issues of expressive queries. Peers use ontologies to describe their shared content. A metric to measure peers' ontology similarity is used to organize peers according to their semantic properties. The network topology is reconfigured with respect to peer's ontological similarity, so that peers with similar ontologies are close to each other. Sunrise [4] is a PDMS that clusters together semantically related peers in a flexible network organization. Each peer is represented by a set of concepts describing its main topics of interest. The network is organized in a set of Semantic Overlay Networks (SONs) [46]. Peers are assigned to one or more SONs based on their exported set of concepts with the help of an *Access Point Structure* (APS), i.e., a centralized structure which maintains a summarized representation of each SON available in the network. Similarity between peers is captured by a distance function considering their exported sets of concepts.

Concerning OntSum, the authors do not describe the ontology matching algorithm used in the system. A simple and asymmetric global measure is used to compute the semantic similarity between two peers' ontologies. In SPEED, we propose a symmetric global measure to determine the similarity between peer schemas (ontologies). The measure is obtained as a result of an ontology matching process which uses linguistic, structural, and semantic matchers. Moreover, OntSum assumes that all peers have the same physical capabilities, which is not common in a P2P system. In SPEED, the physical heterogeneity of participating peers is exploited by designating extra tasks to peers offering better computational capacity. For instance, an integration peer is responsible for integrating query results returned from its data peers and other integration peers. Sunrise concentrates all efforts related to peer clustering in a centralized index structure (APS) which is modified each time a new peer joins or leaves a SON. The frequency of updates in the APS can bring scalability problems to the system. Differently, the semantic index proposed in SPEED is not a structure that fully controls peer connectivity in the system. Instead, its goal is only to provide a promising initial cluster to a requesting peer. Moreover, the frequency of index updates is minimized since summarized cluster ontologies contain the set of most frequent concepts of a cluster which is not constantly modified.

A distinguishing feature of Helios with respect to the other PDMS is that semantic communities are formed in an ad-hoc manner after peers are connected to the system. Such approach enables the formation of dynamic communities since no classification or set of semantic domains needs to be available a priori. However, since the initial neighborhood of peers is defined randomly, unrelated peers may become neighbors in the network. Therefore, during community formation many unrelated peers can be accessed and unnecessary ontology matching comparisons may be executed. The complete absence of any kind of centralized control does not enable a peer to verify the existing communities before starting the formation of a new one. Thus, it is possible that multiple communities dealing with the same topic coexist. To avoid these problems, in SPEED we consider the use of predefined semantic communities. Each community is initially empty and new peers are added to the community as they join the system.

Regarding query reformulation, correspondences in most of these works are restricted to equivalence and subsumption (SomeRDFS also considers disjunction). We go one step further as we also use other kinds of semantic reformulation rules (e.g., closeness and aggregation) which are obtained from the set of semantic correspondences between the peers. When users enable approximation, closeness correspondences may provide expanding concepts related in a given context. Another difference concerns the use we make of disjointness correspondences when there are negations to deal with. We are able to directly obtain the disjoint concept as a solution to the negation of the original concept. Furthermore, the mentioned works do not deal with context. In contrast, our work produces reformulated queries considering the context of the user (preferences), of the query (mode) and of the environment (relevant peers). Moreover, our work prioritizes the generation of exact reformulations. Nevertheless, depending on the context, it also generates an enriched reformulated version, which may avoid empty reformulations, providing a larger set of reformulated queries, and, consequently, non-empty expanded answers to users.

7 Conclusions and Further Work

In highly dynamic environments such as PDMS, the semantics surrounding peers, data, schemas and correspondences is rather important for tasks such as peer clustering or query answering. This work has highlighted the benefits of using such semantics to deal with data management issues in PDMS. We have presented a PDMS, named SPEED, and a semantic-based approach to support peer clustering and query reformulation in the proposed system. To this end, semantic knowledge in the form of ontologies and contextual information are employed in various aspects. Ontologies are used as uniform representation of peers' contents, background knowledge, support for semantic correspondences identification, query answering, and peer clustering. Contextual information is currently used to enhance the query reformulation process. Context-based submitted and reformulated queries may produce different results to different users, depending on the contextual elements that are acquired at their submission time.

Experiments carried out have shown that homogeneous and well-separated clusters are obtained if an incremental semantic-based clustering process is used to organize peers in a PDMS network. This kind of organization also benefits query answering since it allows the system to forward a query only to a small superset of the peers containing relevant data, hence reducing both communication and execution costs. Moreover, considering semantics at query reformulation time allows the system to obtain enriched reformulations substituting empty exact ones (in restricted mode) or adding another reformulation (in expanded mode) as a way of query enrichment. As a result, we obtain a larger set of query reformulation possibilities and users are provided with expanded related answers.

There are a number of ongoing research issues concerned with the use of semantics in PDMSs. Among them, in the near future we intend to focus on two relevant issues: (i) the maintenance of semantic communities; and (ii) query routing. Concerning the former, an issue to be studied in detail regards the evolution of cluster ontologies. In order to reflect the content available in a semantic cluster, cluster ontologies should be created and maintained dynamically, in an automatic way, according to peers' intermittence. A cluster ontology should be able to evolve not only when a requesting peer joins the cluster but also when a participating peer leaves it. The use of a more sophisticated technique to identify a semantic community is also a future work. Regarding query routing, query reformulation strategies and query routing mechanisms [47] have a great influence on each other, and represent the main steps a query answering process in a PDMS must achieve. In our current version, we reformulate queries between two peers and combine answers by means of a union operation. In future work, query routing strategies will be added to improve produced query answers result, i.e., a submitted query will be reformulated in such a way that it is possible to ensure effective query routing, preserving the query semantics at the best possible level of approximation. We intend to use semantics to enhance the selection of relevant semantic neighbors in order to improve query routing. Furthermore, we have been carrying out experiments in order to obtain feedback from users in order to improve query processing tasks and to validate the weights we have arbitrarily defined to the semantic correspondences.

References

1. Halevy, A.Y., Ives, Z.G., Mork, P., Tatarinov, I.: Piazza: Data Management Infrastructure for Semantic Web Applications. In: World Wide Web Conference, pp. 556–567 (2003)
2. Valduriez, P., Pacitti, E.: Data Management in Large-Scale P2P Systems. In: Int. Conference on High Performance Computing for Computational Science (VecPar 2004), Valencia, Spain, pp. 104–118 (2004)
3. Mandreoli, F., Martoglia, R., Penzo, W., Sassatelli, S., Villani, G.: SUNRISE: Exploring PDMS Networks with Semantic Routing Indexes. In: 4th European Semantic Web Conference (ESWC 2007), Innsbruck, Austria (2007)
4. Lodi, S., Penzo, W., Mandreoli, F., Martoglia, R., Sassatelli, S.: Semantic Peer, Here are the Neighbors You Want? In: 11th Extending Database Technology (EDBT 2008), Nantes, France, pp. 26–37 (2008)
5. Kantere, V., Tsoumakos, D., Sellis, T., Roussopoulos, N.: GrouPeer: Dynamic clustering of P2P databases. The Information Systems Journal 34(1), 62–86 (2009)
6. Herschel, S., Heese, R.: Humboldt Discoverer: A Semantic P2P index for PDMS. In: International Workshop Data Integration and the Semantic Web, Porto, Portugal (2005)
7. Castano, S., Montanelli, S.: Semantic Self-Formation of Communities of Peers. In: ESWC Workshop on Ontologies in Peer-to-Peer Communities, Heraklion, Greece (2005)
8. Souza, D., Arruda, T., Salgado, A.C., Tedesco, P., Kedad, Z.: Using Semantics to Enhance Query Reformulation in Dynamic Environments. In: 13th East European Conference on Advances in Databases and Information Systems (ADBIS 2009), Riga, Latvia, pp. 78–92 (2009)
9. Bouquet, P., Kuper, G.M., Scoz, M., Zanobini, S.: Asking and answering semantic queries. In: Workshop on Meaning Coordination and Negotiation Workshop (MCN 2004) in Conjunction with the 3rd International Semantic Web Conference (ISWC 2004), Hiroshima, Japan (2004)
10. Mandreoli, F., Martoglia, R., Villani, G., Penzo, W.: Flexible query answering on graph-modeled data. In: 12th International Conference on Extending Database Technology (EDBT 2009), Saint-Petersburg, Russia, pp. 216–227 (2009)
11. Dey, A.: Understanding and Using Context. Personal and Ubiquitous Computing Journal 5(1), 4–7 (2001)
12. Souza, D., Belian, R., Salgado, A.C., Tedesco, P.: Towards a Context Ontology to Enhance Data Integration Processes. In: 4th Workshop on Ontologies-based Techniques for DataBases in Information Systems and Knowledge Systems (ODBIS), Auckland, New Zealand, pp. 49–56 (2008)

13. Bucur, O., Beaune, P., Boissier, O.: Steps Towards Making Contextualized Decisions: How to Do What You Can, with What You Have, Where You Are. In: Second International Workshop on Modeling and Retrieval of Context (MRC), Edinburgh, Scotland, pp. 62–85 (2005)
14. Bolchini, C., Curino, C., Orsi, G., Quintarelli, E., Rossato, R., Schreiber, F., Tanca, L.: And what can context do for data? Communication of the ACM 52(11), 136–140 (2009)
15. Pires, C.E.: Semantic-based Connectivity in a Peer Data Management System. In: 6th Workshop of Thesis and Dissertation on Data Bases, held in Conjunction with the 22th Brazilian Symposium on Data Bases (SBBD 2008), João Pessoa, Brazil (2007)
16. The Free Network Project (2009), http://freenetproject.org/
17. Stoica, I., Morris, R., Karger, D., Kaashoek, M.F., Balakrishnan, H.: Chord: A Scalable Peer-to-Peer Lookup Service for Internet Applications. In: ACM SIGCOMM, San Diego, California, USA, pp. 149–160 (2001)
18. Yang, B., Garcia-Molina, H.: Designing a Super-Peer Network. In: 19th International Conference on Data Engineering (ICDE 2003), Bangalore, India, pp. 49–60 (2003)
19. Pires, C.E., Souza, D., Kedad, Z., Bouzeghoub, M., Salgado, A.C.: Using Semantics in Peer Data Management Systems. In: Colloquium of Computation: Brazil / INRIA, Cooperations, Advances and Challenges, Bento Gonçalves, Brazil, pp. 176–179 (2009)
20. Pires, C.E., Souza, D., Pachêco, T., Salgado, A.C.: A semantic-based ontology matching process for PDMS. In: Hameurlain, A., Tjoa, A.M. (eds.) Globe 2009. LNCS, vol. 5697, pp. 124–135. Springer, Heidelberg (2009)
21. Pires, C.E., Sousa, P., Kedad, Z., Salgado, A.C.: Summarizing Ontology-based Schemas in PDMS. In: Int. Workshop on Data Engineering meets the Semantic Web (DESWeb 2010) in conjunction with ICDE 2010, Long Beach, CA, USA, pp. 239–244 (2010)
22. Pires, C.E.: Ontology-based Clustering in a Peer Data Management System. PhD Thesis. Federal University of Pernambuco, Recife, Brazil (2009)
23. Euzenat, J., Shvaiko, P.: Ontology Matching. Springer, Heidelberg (2007)
24. Castano, S., Ferrara, A., Montanelli, S.: Matching ontologies in open networked systems: Techniques and applications. Journal on Data Semantics 3870, 25–63 (2006)
25. Borgida, A., Serafini, L.: Distributed description logics: Assimilating information from peer sources. In: Spaccapietra, S., March, S., Aberer, K. (eds.) Journal on Data Semantics I. LNCS, vol. 2800, pp. 153–184. Springer, Heidelberg (2003)
26. Aumüller, D., Do, H.H., Massmann, S., Rahm, E.: Schema and ontology matching with COMA++. In: International Conference on Management of Data (SIGMOD), Software Demonstration (2005)
27. Castano, S., Antonellis, V., Fugini, M.G., Pernici, B.: Conceptual Schema Analysis: Techniques and Applications. ACM Transactions on Database Systems 23(3), 286–333 (1998)
28. Rijsbergen, C.J.: Information Retrieval, 2nd edn. Stoneham, Butterworths (1979), http://www.dcs.gla.ac.uk/Keith/Preface.html
29. Baeza-Yates, R., Ribeiro-Neto, B.: Modern Information Retrieval. ACM Press/Addison-Wesley (1999)
30. Hartigan, J.A.: Clustering Algorithms. John Wiley and Sons, Inc., New York (1975)
31. Fisher, D.H., Xu, L., Zard, N.: Ordering effects in clustering. In: 9th International Conference on Machine Learning, Aberdeen, Scotland, pp. 163–168 (1992)
32. Baader, F., Calvanese, D., McGuinness, D., Nardi, D., Patel-Schneider, P. (eds.): The Description Logic Handbook: Theory, Implementation and Applications. Cambridge University Press, Cambridge (2003)

33. Souza, D.Y.: Using Semantics to Enhance Query Reformulation in Dynamic Distributed Environments. PhD Thesis. Federal University of Pernambuco, Recife, Brazil (2009)
34. Theodoridis, S., Koutroumbas, K.: Pattern Recognition, 2nd edn. Academic Press, London (2003)
35. Batistakis, Y., Halkidi, M., Vazirgiannis, M.: Cluster validity methods: Part I. Sigmod Record 31(12), 40–45 (2002)
36. Fowlkes, E., Mallows, C.: A method for comparing two hierarchical clusterings. Journal of the American Statistical Association 78(383), 569–576 (1983)
37. Batistakis, Y., Halkidi, M., Vazirgiannis, M.: Clustering validity checking methods: Part II. Sigmod Record 31(3), 19–27 (2002)
38. Jain, A.K., Murty, M.N., Flynn, P.J.: Data clustering: a review. ACM Computing Survey 31(3), 264–323 (1999)
39. Rousseeuw, P.J.: Silhouettes: a graphical aid to the interpretation and validation of cluster analysis. Journal of Computational and Applied Mathematics 20, 53–65 (1987)
40. Halevy, A., Ives, Z., Suciu, D., Tatarinov, I.: Schema mediation for large-scale semantic data sharing. VLDB Journal 14(1), 68–83 (2005)
41. Löser, A., Naumann, F., Siberski, W., Nejdl, W., Thaden, U.: Semantic Overlay Clusters within Super-Peer Networks. In: International Workshop on Databases, Information Systems and Peer-to-Peer Computing (DBISP2P 2003), Berlin, Germany, pp. 33–47 (2003)
42. Conforti, G., Ghelli, G., Manghi, P., Sartiani, C.: Scalable Query Dissemination in XPeer. In: IDEAS 2007, pp. 199–207 (2007)
43. Faye, D., Nachouki, G., Valduriez, P.: Semantic Query Routing in SenPeer, a P2P Data Management System. In: Int. Conf. on Network-Based Information Systems (NBiS), Regensburg, Germany, pp. 365–374 (2007)
44. Adjiman, P., Goasdoué, F., Rousset, M.-C.: SOMERDFS in the semantic web. In: Spaccapietra, S., Atzeni, P., Fages, F., Hacid, M.-S., Kifer, M., Mylopoulos, J., Pernici, B., Shvaiko, P., Trujillo, J., Zaihrayeu, I. (eds.) Journal on Data Semantics VIII. LNCS, vol. 4380, pp. 158–181. Springer, Heidelberg (2007)
45. Li, J., Vuong, S.: OntSum: A Semantic Query Routing Scheme in P2P Networks Based on Concise Ontology Indexing. In: 21st International Conference on Advanced Networking and Applications, Niagara Falls, Canada, pp. 94–101 (2007)
46. Crespo, A., Garcia-Molina, H.: Semantic Overlay Networks for P2P Systems. Technical Report, Stanford University (2002)
47. Montanelli, S., Castano, S.: Semantically Routing Queries in Peer-based Systems: the H-Link Approach. Knowledge Eng. Review 23(1), 51–72 (2008)

P2Prec: A P2P Recommendation System for Large-Scale Data Sharing[*]

Fady Draidi[1], Esther Pacitti[1], and Bettina Kemme[2]

[1] INRIA & LIRMM, Montpellier, France
{Fady.Draidi,Esther.Pacitti}@lirmm.fr
[2] McGill University, Montreal, Canada
kemme@mcgill.ca

Abstract. This paper proposes P2Prec, a P2P recommendation overlay that facilitates document sharing for on-line communities. Given a query, the goal of P2PRec is to find *relevant peers* that can recommend documents that are relevant for the query and are of high quality. A document is relevant to a query if it covers the same topics. It is of high quality if relevant peers have rated it highly. P2PRec finds relevant peers through a variety of mechanisms including advanced content-based and collaborative filtering. The topics each peer is interested in are automatically calculated by analyzing the documents the peer holds. Peers become relevant for a topic if they hold a certain number of highly rated documents on this topic. To efficiently disseminate information about peers' topics and relevant peers, we propose new semantic-based gossip protocols. In addition, we propose an efficient query routing algorithm that selects the best peers to recommend documents based on the gossip-view entries and query topics. At the query's initiator, recommendations are selectively chosen based on similarity, rates and popularity or other recommendation criteria. In our experimental evaluation, using the TREC09 dataset, we show that using semantic gossip increases recall by a factor of 1.6 compared to well-known random gossiping. Furthermore, P2Prec has the ability to get reasonable recall with acceptable query processing load and network traffic.

Keywords: P2P systems, recommendation, gossip protocols, decentralization, on-line communities.

1 Introduction

The context of this work is Peer-to-Peer (P2P) on-line communities, where users want to share content at very large scale. For instance, in modern e-science, such as bio-informatics, physics and environmental science, scientists must deal with overwhelming amounts of contents (experimental data, documents, images, etc.). Such content is usually produced and stored locally in the workspace of individual users. Users are often willing to share their content with the community or with specific colleagues or friends if they can, in return, receive content from others. Our

[*] Work partially funded by the DataRing project of the French ANR.

A. Hameurlain, J. Küng, and R. Wagner (Eds.): TLDKS III , LNCS 6790, pp. 87–116, 2011.

goal is to develop a system that supports such sharing without a central server, which is often difficult and expensive to maintain.

P2P networks offer scalability, dynamicity, autonomy and decentralized control. These are the properties that we want to exploit. P2P-based file-sharing systems have proven very efficient at locating content given specific queries. However, few solutions exist that are able to recommend the most relevant documents given a keyword based query. In this paper, we address this problem with P2Prec, a recommendation system for P2P on-line communities.

Most recommender systems for web data are centralized and are either content-based or use collaborative filtering. Content-based systems recommend to a user u items that are similar to u's previously rated items. Collaborative filtering, in contrast, recommends to u items that have been rated by users who share similar interests based on the tagging or rating behavior. Typically, recommendation is done using a matrix model [1] that is known to be space consuming and suffers from sparsity (the fact that many users only rate a small numbers of items) and limited scalability.

Decentralized recommendation for web data based on collaborative filtering has been recently proposed [20] with promising results. But there is much room for improvement. Our approach leverages content-based and collaborative filtering recommendation approaches. In most collaborative filtering systems, topics of interest are derived based on the users' tagging activities that may lead to ambiguous interpretations. In contrast, in our context of online communities, we exploit the fact that people tend to store high quality content related to their topics of interests. Thus, we can automatically derive the users' topics of interest from the documents they store and the ratings they give, without requiring tagging.

P2Prec works with a set of documents distributed over a large-scale network of volunteer and autonomous peers (users) willing to share and rate their documents. It automatically extracts the topics a user is interested in by relying on a generic automatic topic classifier such as Latent Dirichlet Allocation (LDA) [5]. LDA uses Bayesian statistics and machine learning techniques to infer to the hidden topics in unlabeled content (documents, collections of images, music, DNA sequences, etc) from labeled content whose topics have already been determined.

To guide recommendation and manage sparsity,, we propose a metric to identify the relevance of a user with respect to a given topic. That is, a user is considered *relevant* to give recommendations for a specific topic t if it has a sufficient number of highly rated documents related to t.

Information about the interest and relevance of users is disseminated over the P2Prec overlay using gossip protocols. With *random gossiping* [12, 17], each user keeps locally a view of its dynamic acquaintances (or view entries), and their corresponding topics of interest. Periodically, each user chooses randomly a contact (view entry) to gossip with. The two involved peers then exchange a subset of each other's view, and update their view state. This allows peers to get to know new peers and to forget about peers that have left P2Prec. Whenever a user submits a query, the view is used as a directory to redirect the query to the appropriate peers. Thus, overlay maintenance and information dissemination are done gracefully, assuring load balancing and scalability. Several algorithm parameters, such as the gossip contact, the view subset, etc. are chosen randomly.

In P2Prec, users search for documents that are related to their topics of interests. Thus, in order to increase the quality and the efficiency of recommendation, we propose a *semantic gossip* approach where semantic information, such as user's topics of interest, is taken into account while gossiping.

In this paper, we make the following contributions:

1. We propose a new approach for decentralized recommendation that leverages collaborative filtering and content-based recommendation. To guide recommendation, we introduce the concept of *relevant users*.
2. We propose a P2Prec overlay that enables efficient decentralized recommendation using gossip protocols. We propose two new semantic-based gossip algorithms that take into account semantic information such as the users' topics of interest and user relevance, while maintaining the nice properties of gossiping.
3. We propose an efficient query routing algorithm that takes into account the most relevant view entries, and recommends the best users to provide recommendation for a query. We use information retrieval techniques, such as *cosine similarity,* to help P2Prec find relevant documents at each involved peer.
4. To rank recommendations at the query initiator, we propose a rank method that takes into account similarities, ratings and document popularity.
5. We provide an experimental evaluation using the TREC09 dataset [29] that demonstrates the efficiency of P2Prec.

The rest of this paper is organized as follows. Section 2 defines the problem. Section 3 introduces P2Prec basic concepts such as topics of interest and relevant users. Section 4 describes how the P2Prec overlay is constructed and maintained via gossip protocols. Sections 5-6 describe two new gossip algorithms, semantic and semantic two-layered gossip, respectively. Section 7 describes our solution for query routing and result ranking. Section 8 gives an experimental evaluation. Section 9 discusses related work. Section 10 concludes.

2 Problem Definition

Intuitively, given a query, we want to recommend the most relevant and qualitative documents from a huge distributed content base. Most recommender systems (RS) for web data are centralized and are either content-based or use collaborative filtering. Content-based systems recommend to a user u items that are similar to u's previously rated items. Typically, they measure the similarity between the items u has rated and all rated items in the system, and then recommend to u items with high similarity. Collaborative filtering (CF), in contrast, recommends to u items that have been rated by users who share similar interests with u. Typically, CF first measures the similarity between u and all users in the system based on their rating or tagging. Then, it selects those users who are most similar to u. Finally, CF normalizes and computes the weighted sum of the selected users' ratings, then makes recommendations based on those ratings. Relying on both content-based and collaborative filtering approach, we extract a user's topics of interest based on the documents stored at the user.

Our recommendation model assumes a set D of shared documents and a set U of users $u_1,...u_n$ corresponding to autonomous peers $p_1,...p_n$. Notice that documents may be replicated as a result of using P2Prec. Thus, each document doc_i can have many read-only copies. Since we focus on on-line communities, we safely assume that users are willing to rate the documents they store. That is, each document doc_i that has a copy at user u has high probability to be rated by u. Furthermore, we assume a set T of topics. Our system will automatically associate each user $u \in U$ with a set of topics of interest $T_u \subset T$, and a set of relevant topics $T_u^r \subset T_u$ depending on the documents u maintains locally and the ratings he/she has given to these documents. More specifically, a topic t is of interest for user u, i.e, $t \in T_u$, if a specific percentage of u's local documents D_u are related to topic t with high probability and are highly rated by u. User u is considered a *relevant user* for topic $t \in T_u^r \subset T_u$, if u is interested in t and has a sufficient number of highly rated documents that are related to t with high probability, and u will be able to provide high quality recommendations related to t.

Finally, queries are expressed through key-words and a response to a query q is a recommendation defined as:

$$recommendation_q = rank(rec_q^{1}(doc_i), \ldots rec_q^{n}(doc_j))$$

Different recommendations $rec_q^{1}(doc_i)$, $rec_q^{2}(doc_i)$, ... may be given for the replicas of a *document doc_i*. Each *rec* is defined in terms of the similarity between the query *q and doc_i*, and the document popularity. Finally, the rank function may be standard or user defined.

Problem Statement: Given a key-word query q and our recommendation model above, the problem we address is how to efficiently retrieve the most relevant users (or peers) to compute *recommendation$_q$* and selectively choose the best recommendations.

3   P2Prec Basic Concepts

In this section, we introduce P2Prec basic concepts for managing topics of interests and relevant users. First we present how topics are extracted from a set of documents by using LDA. Next, we introduce how we extract users' topics of interests from documents they store, and how we define the concept of relevant users.

3.1   Topics Extraction

In P2Prec, topics are automatically extracted from a set of documents to produce the set of topics T, and for each user $u \in U$, its set of topics of interest $T_u \subset T$. Classifying the hidden topics available in a set of documents is an interesting problem by itself. Several models have been proposed, described and analyzed in the Information Retrieval (IR) literature [7] to tackle this problem. We use, LDA, a topic classifier model that represents each document as a mixture of various topics and models each topic as a probability distribution over the set of words in the document. For example, a document talking about vegetarian cuisine is likely to be generated from a mixture of words from the topics food and cooking.

We adapt LDA for P2Prec to proceed in two steps: the training (at the global level, see Figure 1(a)), and inference (at the local level, see Figure 1(b)). Training is usually done by a specific peer, e.g., the bootstrap server. LDA is fed with a sample set of M documents that have been aggregated from the system, i.e., collected from

P2Prec peers on demand. Each document $doc \in M$ is a series of words, $doc=\{word_1,...,word_n\}$, where $word_i$ is the ith word in doc and n is the total number of words in doc. Then, LDA executes its topic classifier program and produces a set $B=\{b_1,... b_d\}$ of bags. Each bag $b \in B$ is tagged with a label t (we refer to it as topic t). The set of topics T of P2Prec corresponds to $t_1...t_d$. Each bag contains a set of z words, where z is the total number of the unique words in M, and each of these words is associated with a weight value between 0 and 1. More formally, this set of bags can be represented as a matrix ϕ with dimensions $d*z$, where d is the number of topics and z is the total number of unique words in M. Each row of ϕ represents the probability distribution of a topic $t \in T$ over all words. The bootstrap server periodically aggregates M from the peers and estimates ϕ. Each version of ϕ is attached with a timestamp value.

The inference part of LDA is performed locally at each (peer) user u. The goal is to extract the topics of $u's$ local documents, using the same set of topics that were previously generated at the global level. Thus, whenever a peer joins P2Prec, it first contacts the bootstrap server in order to download ϕ. Then, for inference, LDA's input is the set of local documents of user u, and the matrix ϕ generated at the global level. As output, LDA produces a vector of size d for each document doc, called document topic vector, $V_{doc}=[w_{doc}^{t_1}....w_{doc}^{t_d}]$, where $w_{doc}^{t_i}$ is the weight of each topic $t \in T$ with respect to doc.

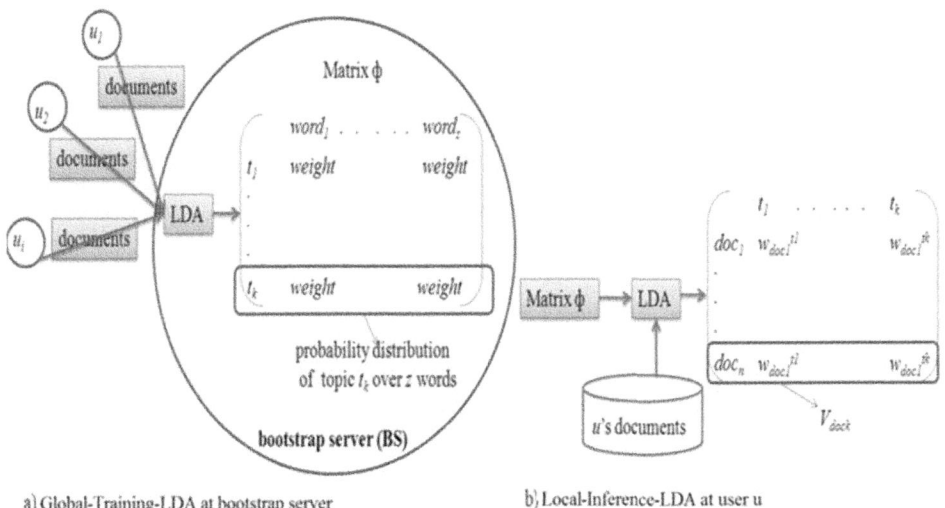

a) Global-Training-LDA at bootstrap server   b) Local-Inference-LDA at user u

Fig. 1. LDA under P2Prec context

3.2   Topics of Interest and Relevant Users

Now introduce the concepts of users' topics of interest and relevant users necessary to guide recommendation. Algorithm 1 illustrates how each user computes its topics of interests with which we determine relevant users. Given a set of documents D_u stored locally at user u, we extract the topics of interest $T_u \subset T$ in two steps. First, we compute the document quality for each document $doc \in D_u$ that user u has rated and we record the quality locally in a vector $quality(doc,u)$. This is done by multiplying the document topic vector $V_{doc}=[w_{doc}^{t_1} \dots w_{doc}^{t_d}]$ that has been extracted using the **Local-Inference-LDA**, by the rate $rate_{doc}^u$ that has been given by user u for doc. Thus, we have:

$$quality(doc,u) = [w_{doc}^{t_1} * rate_{doc}^u \dots w_{doc}^{t_d} * rate_{doc}^u] \text{ (corresponds to line 2)}$$

Then, user u identifies for each topic $t \in T$ only the documents that are highly related to t. A document doc is considered highly related to topic t, denoted by $relate_t(doc, u)$, if its weight in that topic w_{doc}^t multiplied by its rate $rate_{doc}^u$ exceeds a threshold value (which is system defined), i.e.,

$$relate_t(doc,u) = \begin{cases} 1, & w_{doc}^t * rate_{doc}^u \geq threshold \\ 0, & otherwise \end{cases}$$

Topic(s) related to a document leverages the user's rating and document's semantic content. If a document has not been rated explicitly by user u, we still have the ability to compute the topics that are related to it. In this case, we consider the document to be highly related to topic t if its weight in that topic exceeds a threshold value.

In the second step (lines 3, 4 and 5), user u counts how many documents are highly related to each topic $t \in T$. The number of documents that are highly related to $t \in T$ represents u's degree of interest in that topic, denoted by $degree_u^t$, i.e.,

$$degree_u^t = \sum_{i=1}^{|D_u|} relate_t(doc_i, u)$$

Then, user u implicitly computes its topics of interest $T_u \subset T$ (lines 9, 10 and 11). User u is considered interested in topic t if a percentage y (or absolute value) of its local documents D_u are highly related to topic t, i.e.,

$$t \in T_u \ if \ \frac{degree_u^t}{|D_u|} \geq y$$

Furthermore, user u is considered a relevant user for topic t that belongs to its relevant topics T_u^r i.e., $t \in T_u^r$, if u is interested in t and $degree_u^t$ exceeds a number x (which is system defined), i.e.,

$$t \in T_u^r \ if \ degree_u^t \geq x \ and \ t \in T_u$$

In other words, u is considered a relevant user in topic $t \in T_u^r$ if it is interested in t and has a sufficient amount of documents that are highly related to topic $t \in T_u^r$ (lines 12 and 13). Otherwise u is not a relevant user for topic t even though u might be interested in t. We denote a user that is not relevant for any topic as a *non-relevant user*.

Algorithm 1- Compute-Topics-Of-Interest(V_{doc}, $rate_{doc}$)
Input: user u's document topic vectors, V_{doc} where $doc \in D_u$; user u's document rating, $rate_{doc}^u$ where $doc \in D_u$
Output: user u's topics of interest T_u
1   **For** each $doc \in D_u$ **do**

```
2        quality(doc,u) = Multiply(V_doc, rate_doc^u)
3        For each t∈T do
4            If relate_t(doc,u) then
5                increase degree_u^t by one
6            End If
7        End For
8    End For
9    For each t∈T do
10       If (degree_u^t / |D_u|) ≥ y  then
11           u add t to T_u
12           If degree_u^t ≥ x  and t∈T_u then
13               u add  t to T_u^r
14           End If
15       End If
16   End For
```

User u has the ability to download and rate the documents it receives, and add or delete documents. Thus, its relevance (topics of interest) may change over time. To capture this dynamic behavior, user u computes its topics of interest T_u and relevant topics T_u^r periodically, or if a number of documents have been added to (or deleted from) its D_u and exceeds a system-defined threshold.

4 P2Prec Overlay

In this section, we first describe how the P2Prec overlay is constructed and maintained via gossip algorithms. Then, we introduce our query routing solution. Finally, we describe the well-known random gossip algorithm [12, 17], and discuss its limitations for P2Prec.

4.1 Overlay Construction

In P2Prec, users use gossip-style communication to construct the P2Prec overlay and exchange a subset of their views in an epidemic manner [14]. Users also gossip to detect failed users. Gossip algorithms [12, 17] have attracted a lot of interest for building and managing unstructured networks. With gossip, each user periodically (with a gossip period noted C_{gossip}) exchanges a subset of its state, called *local-view*, with another user. Thus, after a while, as with gossiping in real life, each user will have a partial view of the other users in the system.

Each user's *local-view* contains a fixed number of entries, noted *view-size*. Each entry refers to a user u, and contains u's gossip information such as:

- u's IP address;
- u's topics of interest T_u, each topic $t∈T_u$ being associated with a Boolean field that indicates whether u is relevant in that topic.

Users' topics of interest and relevant users' information are disseminated using gossip algorithms in order to guide queries for recommendation retrieval. When a user is interested in a topic t it may be a candidate to serve a query on topic t. This corresponds to the case in which there is no relevant user on that topic in the view.

In the example of Figure 2, u carries in its *local-view* two users v_1 and v_2. User v_1 is interested in two topics t_1 and t_2. Figure 2 shows that v_1 is not relevant either in t_1 or t_2. As user v_1 is not relevant in any topic, then v_1 is a non-relevant user. User v_2 is interested in two topics t_1 and t_2. Figure 2 shows that v_2 is relevant in t_2, and not relevant in t_1. As v_1 is relevant at least in one topic, then v_2 is a relevant user.

$$T_u = <t_1, t_2>$$

user	T_{vi}
v_1	$(t_1, 0); (t_2, 0)$
v_2	$(t_1, 0); (t_2, 1)$

Fig. 2. User u's *local-view*

We choose gossip-style communication for the following reasons. First, the continuous exchange of subsets of *local-views* between users enables the building of an unstructured overlay network in a continuous manner, which reflects the natural dynamism of P2P networks and helps providing very good connectivity despite failures or peer disconnections [12]. Second, gossiping provides a reliable way to disseminate information in large-scale dynamic networks, so that users discover new users [19]. Third, it ensures load balancing during the disseminating of information between users, since all users have the same number of gossip targets and the same exchange period, and thus send exactly the same number of messages [12]. Finally, it is scalable, reliable, efficient and easy to deploy [16].

4.2 Query Processing

Whenever a user submits a query, *local-view* is used as a directory to redirect the query to the appropriate relevant users. A query is defined as $q(word_i, TTL, V_q, T_q, u)$, where $word_i$ is a list of keywords, TTL is the time-to-live value, and V_q is query q's topic vector. Notice that q's topic vector V_q is computed using the **Local-Inference-LDA**. T_q represent q's topics and u corresponds to the address of q's initiator. When a user u initiates a query q, it routes q as follows: first, it extracts q's topic vector V_q using the **Local-Inference-LDA**. Then, user u computes q's topics T_q from q's topic vector V_q. The query q is considered to belong to a topic $t \in T_q$ if its weight w_q^t in that topic exceeds a certain threshold (which is system defined). Then, user u uses its *local-view* to find relevant users that can give recommendation for q's topics T_q, and then redirects q to those relevant users after reducing TTL.

Whenever a user u receives a query q that has been initiated by a user v, it returns to q's initiator the recommendation information it has which are related to q, and recursively selects from its *local-view* the relevant users in q' topics T_q. Afterwards, u redirects q to those relevant users as long as TTL does not reach zero. More details on query processing are given in Section 7.

4.3 Random Gossip Algorithm

The basic random gossip algorithm (Rand for short) proceeds as follows: a user u (either relevant or non-relevant) acquires its initial *local-view* during the join process using a bootstrap technique. We register each user that has joined P2Prec at a bootstrap server. Whenever a user u joins the system, it randomly selects a set of users from the bootstrap server to initialize its *local-view*. Notice that u's *local-view* may carry relevant and non-relevant users.

Whenever a user u initiates an information exchange, it selects a random contact v from its *local-view* to gossip with. Then, u selects a random subset of size L_{gossip} -1, noted *viewSubset*, from its *local-view*, and includes itself into *viewSubset*. Then, u sends *viewSubset* to v. Similarly, u receives a *viewSubset** of v's *local-view*.

Finally, once a user u receives a gossip message, it updates its *local-view* based on the gossip message received. The update process proceeds as follows: 1) the content of the gossip message is merged with the content of the current *local-view* of user u and set in a buffer; 2) using the buffer, u selects *view-size* entries randomly and updates its *local-view*. Whenever, u searches for a recommendation, it uses its *local-view* to identify the relevant users in its view that can provide recommendation for the query.

Rand does not take into account user u's topics of interests during the gossip exchanges. This reduces the possibility of having users in u's *local-view* which are similar to u in terms of topics of interest, which reduces the possibility of getting better responses. In particular, the exchange does not consider whether the view contains users that are relevant for the topics user u is interested in. In the following we refer to this as similarity. For now, we informally assume two users are similar if they have similar interests. In Section 5.2 we provide a formal definition.

In the following we present three examples where Rand limits the quality of the gossip exchange:

1. Consider two users u and v_1 that are not similar, and user v_1 is in u's *local-view* (see Figure 3(a)), because they do not have any common topic of interest. Let us suppose that v_1 has several users in its *local-view* that are similar to v_1. For instance v_4 is in v_1's *local-view*, and has topics of interest T_{v4} which are the same as v_1 topics of interest T_{v1}. By transitivity these users are not similar to u. If u chooses v_1 as a gossip contact, with high probability it will end by filling its *local-view* with un-similar users (see Figure 3(b)), because most of the users in u's *local-view* do not have topic in common with u's topics of interest. In the example of Figure 3, u selects v_1 to gossip with, and sends to v_1 *ViewSubset$_u$* which, in addition to itself, includes v_2 and v_3. Similarly, v_1 returns to u a *vewSubset$_{v1}$* which includes in addition to itself users v_5 and v_6. Once u receives the *viewSubset$_{v1}$*, it merges *viewSubset$_{v1}$* with its *local-view* in a buffer, and then updates its *local-view*.

$T_u = <t_1,t_2>$		$T_{v1} = <t_3,t_4>$	
user	T_{vi}	user	topics
v_1	$(t_3,1);(t_4,1)$	v_4	$(t_3,1);(t_4,1)$
v_2	$(t_1,1);(t_2,1)$	v_5	$(t_3,1);(t_5,0)$
v_3	$(t_2,1);(t_3,0)$	v_6	$(t_4,1)$

$T_u = <t_1,t_2>$		$T_{v1} = <t_3,t_4>$	
user	T_{vi}	user	topics
v_5	$(t_3,1);(t_5,0)$	u	$(t_1,1);(t_2,1)$
v_2	$(t_1,1);(t_2,1)$	v_2	$(t_1,1);(t_2,1)$
v_6	$(t_4,1)$	v_6	$(t_4,1)$

a) u and v_1 local-views before gossip   b) u and v_1 local-views after gossip

Fig. 3. Users u and v are not similar

2. Consider now that u and v are similar (see Figure 4(a)), because u's topics of interest T_u and v_1's topics of interest T_{v1} are similar. However, v_1 has many un-similar users in its *local-view*. For instance v_4 is in v_1's *local-view* and does not have any topic of interest that v_1 is interested in. By transitivity, these users are not similar to u. If u chooses v as gossip contact again, with high probability it will end up filling its *local-view* with un-similar users (see Figure 4(b)), because most of the users in u's *local-view* do not have topic in common with u's topics of interest. In the example of Figure 4, u selects v_1 to gossip with, and sends to v_1 ViewSubset$_u$ which includes in addition to itself users v_2 and v_3. Similarly, v_1 returns to u a vewSubset$_{v1}$ which includes in addition to itself users v_5 and v_6. Once u receives the viewSubset$_{v1}$, it merges viewSubset$_{v1}$ with its *local-view* in a buffer, and then updates its *local-view*.

3. Consider the case that several users $u_1,...,u_k$ are non-relevant users, and u's *local-view* carries mostly non-relevant users (see Figure 5(a)), for example v_2 is in u's *local-view* and is not relevant in any topic. In this case, the gossip exchanges are useless for serving queries (see Figure 5(b)). For example, after gossiping, u does not carry in its *local-view* any user that is relevant in topic $t_1 \in T_u$.

$T_u = <t_1,t_2>$		$T_{v1} = <t_1,t_2>$	
user	T_{vi}	user	topics
v_1	$(t_1,1);(t_2,1)$	v_4	$(t_3,1);(t_4,1)$
v_2	$(t_1,1);(t_2,1)$	v_5	$(t_3,1);(t_5,0)$
v_3	$(t_2,1);(t_3,0)$	v_6	$(t_4,1)$

$T_u = <t_1,t_2>$		$T_{v1} = <t_1,t_2>$	
user	T_{vi}	user	topics
v_5	$(t_3,1);(t_5,0)$	v_4	$(t_1,1);(t_2,1)$
v_6	$(t_4,1)$	v_3	$(t_2,1);(t_3,0)$
v_3	$(t_2,1);(t_3,0)$	v_6	$(t_4,1)$

a) u and v_1 local-views before gossip   b) u and v_1 local-views after gossip

Fig. 4. User u and v are similar

$T_u = <t_1,t_2>$		$T_{v1} = <t_1,t_2>$	
user	T_{vi}	user	topics
v_1	$(t_1,1);(t_2,0)$	v_4	$(t_1,1);(t_2,0)$
v_2	$(t_1,0);(t_2,0)$	v_5	$(t_2,1);(t_3,0)$
v_3	$(t_2,1);(t_3,0)$	v_6	$(t_1,0)$

$T_u = <t_1,t_2>$		$T_{v1} = <t_1,t_2>$	
user	T_{vi}	user	topics
v_5	$(t_2,1);(t_3,0)$	v_4	$(t_1,1);(t_2,0)$
v_6	$(t_1,0)$	v_2	$(t_1,0);(t_2,0)$
v_3	$(t_2,1);(t_3,0)$	v_6	$(t_1,0)$

a) u and v_1 local-views before gossip   b) u and v_1 local-views after gossip

Fig. 5. User u and v carry mostly non-relevant users

Based on the above examples, we conclude that Rand may generate uninteresting view states resulting in low query responses.

5 Semantic Gossiping

In this section, as a first approach to Rand's limitations, we present a new semantic gossip algorithm (called Semt). The goal is to selectively maximize the number of relevant users at each user u's *local-view* that are similar to u. First, we give our criteria for keeping similar relevant users in the *local-views*. Then, we present in detail the active and passive behavior of Semt.

Recall that our objective is to improve the efficiency of returning useful recommendations for on-line communities. We let each user u maintain a *local-view* of relevant users similar to u. Thus, when u initiates a query q (see Algorithm 3), it searches for a relevant user $v \in u$'s *local-view* so that v can give recommendation for q. If u finds such relevant user, then u's *hit-ratio* is increased. *Hit-ratio* is defined as the percentage of the number of queries that have been answered. Moreover, u likes to find many relevant users in its *local-view* that can serve its queries, and this reduces query response time (time spent to retrieve useful recommendations).

To measure user u's *hit-ratio*, we use a *query-history* that keeps the track of past queries. With Semt, when a user u chooses a contact, it selects a user v that has high *hit-ratio*, and is similar to user u. For that, u includes into *viewSubset* the relevant users that are similar to u, and have high *hit-ratios*. Note that *hit-ratio* can be easily added as an attribute of a *local-view* entry, and becomes part of the gossip message.

In the rest of this section, we present our techniques to compute *hit-ratio*, and the similarity functions.

5.1 Computing the Hit-Ratio

To compute a user's *hit-ratio*, we assume that each user u maintains a log of limited size, called *query-history*, denoted by H_u. The cardinality of u's *query-history* is denoted by $|H_u|$. H_u contains a set of entries, each entry referring to a past query q that u has initiated. Each past query q entry included in H_u contains q's topics T_q and its query state s_q. Query state s_q can be either 1 or 0. The value of 1 for s_q denotes a *query-success*, i.e., there was at least one relevant user in u's *local-view* that was able to serve query q. In contrast, $s_q = 0$ denotes a *query-fail*, i.e., user u has not found any relevant user in its *local-view* that can give recommendations for query q. We use FIFO to replace the past queries once user u's *query-history* has reached its full size $|H_u|$.

Periodically, each user u computes its *hit-ratio*. User u's *hit-ratio* represents the percentage of the number of *query-success* in its *query-history* H_u which is:

$$hit\text{-}ratio_u = \frac{\sum_{i=1}^{n} s_{q_i} \in H_u}{|H_u|} \qquad if\ s_{q_i} = 1$$

where n is the total number of past queries available at u's *query-history* H_u.

5.2 Similarity Functions

Recall that each user has a set of topics of interest, and each relevant user v a set of relevant topics. Thus, we measure the similarity between a user u and a relevant user v, denoted by *distant(u,v)*, by counting the overlap between u's topic of interests T_u and v's relevant topics T_v^r. We use the Dice coefficient [9] which is:

$$distant(u,v) = \frac{2|T_u \cap T_v^r|}{|T_u|+|T_v^r|}$$

We could also use other similarity functions such as cosine, jaccard, etc. Similarly, we use the Dice coefficient [9] to measure the similarity between a query q and a relevant user v:

$$distant(q,v) = \frac{2|T_q \cap T_v^r|}{|T_q|+|T_v^r|}$$

If *distant(q,v)* $\neq 0$, then the relevant user v can give recommendations for q.

5.3 Semantic Gossip Behaviors

The behavior of Semt at a user u is illustrated in Algorithm 2. The active behavior describes how u initiates a periodic gossip exchange message, while the passive behavior shows how u reacts to a gossip exchange initiated by some other user v. Each user u acquires its initial *local-view* during the join process using a bootstrap technique. We register each relevant user which has joined the P2Prec at a bootstrap server. Whenever a user u joins the system, it selects randomly a set of relevant users from the bootstrap server to initialize its *local-view*. Notice that u's *local-view* only carries relevant users.

The active behavior is executed every time unit C_{gossip}. A user u initiates a communication message and computes the similarity distance between itself and each relevant user v in its *local-view* (line 4). Then, u computes the rank of each relevant user v in its *local-view*, denoted by *rank(v)*. A relevant user v's rank at user u depends on the similarity distance between u and v, and v's *hit-ratio* if v has issued more than z number (where z is system defined) of queries within an interval of time i.e., $H_v \geq z$. Otherwise v's rank depends on similarity distance between u and v only. Accordingly the *rank(v)* is:

$$rank(v) = \begin{cases} distant(u,v) & if\ |H_v| \leq z \\ hit\text{-}ratio_v * distant(u,v) & otherwise \end{cases}$$

Usually z is very small, that is to prevent the relevant users that are similar to u, but do not issue queries from getting very low ranks. Note that $|H_v|$ can be easily added as an attribute of a *local-view* entry, and becomes part of the gossip message.

Once u has computed the rank of each relevant user v *rank(v)* in its *local-view*, adds *rank(v)* to a *RankList* (lines 5 to 10) which contains the relevant users' entries along with their ranks. Once u has computed the relevant users' ranks and added them in the *RankList*, it selects from the *RankList* a relevant user v which has the highest rank to gossip with, using the **selectTop()** method (line 12). The relevant user v with the highest rank is the relevant user that is most similar to u and has the highest *hit-ratio_v*.

Once user u has selected a relevant user v to gossip with, it selects L_{gossip} entries from the *RankList* which have the highest rank using **SelectTopEntries()** (line 13). These entries compose user u *viewSubset*. After that user u sends to v the *viewSubset* (line 14).

In turn, user u will receive a *viewSubset** of user v's *local-view* (line 15). Upon receiving *viewSubset**, u computes the rank for each relevant user v in *viewSubset** and adds it to the *RankList* (lines 16-24). Recall that *RankList* includes also the rank of the relevant users at u's current *local-view*. Then, the method **SelectTopEntries()** selects *view-size* entries from the *RankList* which have the highest rank to become the new *local-view* (line 25).

In the passive behavior, a user u waits for a gossip message from a user v. Upon receiving a message (line 3), it computes the rank of the relevant users in its *local-view* (lines 4-11). Then, it uses **SelectTopEntries()** to select *viewSubset** of L_{gossip} entries from the *RankList* that have the highest rank (line 12). Then, it sends back *viewSubset** to user v. Then, it computes the rank of the relevant users in the received *viewSubset* (lines 14-22). Finally, it updates its *local-view* by selecting *view-size* entries from the *RankList* that have the highest rank.

Letting each user u select the top ranked entry v from its *local-view* as the next gossip contact may deteriorate the randomness of its *local-view* entries, because it may occur that v remains the same contact for long period of time. To increase the randomness and prevent user u from selecting the same contact v for a long period of time, each user u stores in a list L, the last / recent contacts that have been selected for gossiping. Then, instead of blindly selecting the top ranked relevant user in *RankList*, to gossip with it selects the first user in *RankList* that is not in the list L of users with whom u has recently gossiped.

Furthermore, the fact that *viewSubset* and the gossip contact are not chosen randomly may reduce the user's ability to discover new relevant users. To overcome this limitation, we propose a semantic two-layered gossiping (Section 6).

Algorithm 2- **Gossiping(***local-view$_u$***)**
//Active behavior
Input: *local-view$_u$*
Output: updated *local-view$_u$*

```
1   Forever do
2      wait(C_gossip)
3      For each relevant user v∈ local-view_u do
4         user u computes distant(u,v)
5         If |H_v| ≥ z then
6            rank(v) = distant(u,v)
7         Else
8            rank(v) = hit-ratio_v * distant(u,v)
9         End If
10        user u adds <rank(v) ,v> to RankList
11     End For
12     user v = selectTop(RankList)
13     viewSubset = SelectTopEntries(RankList,L_goosip)
```

14   User u **send** <viewSubset > to user v
15   User u **receive** viewSubset * from user v
16   **For** each relevant user $v \in$ viewSubset * **do**
17   user u **computes** distant(u,v)
18   **If** $|H_v| \geq z$ **then**
19   rank(v) = distant(u,v)
20   **Else**
21   rank(v) = hit-ratio$_v$ * distant(u,v)
22   **End If**
23   user u **adds** <rank(v) ,v> to RankList
24   **End For**
25   Local-view$_u$ =**SelectTopEntries**(RankList, view-size)
//Passive behavior
Input: viewSubset of a user v; local-view$_u$
Output: updated local-view$_u$
1   **Forever do**
2   **waitGossipMessage( )**
3   **receive** <viewSubset > from user v
4   **For** each relevant user $v \in u's$ local-view **do**
5   user u **computes** distant(u,v)
6   **If** $|Hv| \geq z$ **then**
7   rank(v) = distant(u,v)
8   **Else**
9   rank(v) = hit-ratio$_v$ * distant(u,v)
10   **End If**
11   **End For**
12   viewSubset* = **SelectTopEntries**(RankList,L_{goosip})
13   **send** viewSubset* to user v
14   **For** each relevant user $v \in$ viewSubset **do**
15   user u **computes** distant(u,v)
16   **If** $|Hv| \geq z$ **then**
17   rank(v) = distant(u,v)
18   **Else**
19   rank(v) = hit-ratio$_v$ * distant(u,v)
20   **End If**
21   user u **adds** <rank(v) ,v> to RankList
22   **End For**
23   Local-view$_u$ =**SelectTopEntries**(RankList, view-size)

6   Semantic Two-Layered Gossiping

In this section, we propose a semantic two-layered gossiping (called 2LG) to combine the benefits of Rand (e.g., connected overlay, ability to find new users, etc.) and semantic exchange. Rand preserves gossiping properties and gives users the ability to discover new relevant users. These new relevant users are then taken into account in Semt to find new similar relevant users.

2LG uses the following approach. Each user u maintains a view for each algorithm: 1) a view for Rand, called *random-view* (first layer), with limited size R_{size} , 2) a view for Semt, called *semantic-view* (second layer), with limited size S_{size} s.t. $R_{size} > S_{size}$. Notice that user u uses both Rand and Semt views to support its queries.

With 2LG, each user u acquires its initial *random-view* during the join process (as described in Section 4.3). Then, it initializes its *semantic-view* by computing the ranks of the relevant users in its initial *random-view* and selects S_{size} entries which have the highest ranks. Then, u periodically (with a gossip period C_{random} and $C_{semantic}$) performs Rand and Semt asynchronously. Notice that $C_{semantic} >> T_{random}$ because user semantics (topic of interests) are not changed rapidly. But we assume that C_{random} is small enough to capture the dynamicity of the network, as peer joins and leaves keep happening continuously.

In 2LG, we adopt Semt (see Algorithm 2) with a modification to its active behavior only, to take advantage of the *random-view*. Figure 6 shows the modifications on the active behavior of Semt for 2LG. In principle, the lines 1-24 of Algorithm 2 do not change except that user u uses its *semantic-view* and not the *local-view* for creating the *RankList*. Thus, these lines are not repeated in Figure 5. However, line 25 of Algorithm 2 is replaced by the steps taken in Figure 6. After line 16 of Algorithm 2, the *RankList* includes the rank of the relevant users at u's current *semantic-view* and the ranks of the relevant users in the *viewSubset* that u has received during the exchange. From there, and different to Semt, 2LG also takes into account the relevant users in its *random-view* as follows: u ranks the relevant users in its *random-view*, and adds them to the *RankList* (lines 1-9 in Figure 6). Then, u selects the S_{size} entries from *RankList* that have the highest rank to be its new *semantic-view* (line 10 Figure 6).

Input: *semantic-view$_u$*; *random-view$_u$*
Output: updated *semantic-view$_u$*
Lines 1-24 of the active behavior of Algorithm 2
1   **For** each relevant user $v \in random\text{-}view_u$ **do**
2   user u **computes** $distant(u,v)$
3   **If** $|H_v| \geq z$ **then**
4            $rank(v) = distant(u,v)$
5   **Else**
6            $rank(v) = hit\text{-}ratio_v * distant(u,v)$
7   **End If**
8   user u **adds** $<rank(v), v>$ to *RankList*
9   **End For**
10   *semantic-view$_u$* =**SelectTopEntries**(*RankList*, S_{size})

Fig. 6. The modifications on the active behavior of Semt for 2LG

In the example of Figure 7, we show the framework of 2LG at user u. User u performs Rand and Semt asynchronously. It performs Rand as described in Section 4.3: it selects randomly a user v_1 from its *random-view* to gossip. Then it selects randomly a *viewSubset_u* from its *random-view* and sends it to v_1. Afterwards, user *u* receives a *viewSubset_{v1}* from v_1. Once *u* has received *viewSubset_{v1}*, it updates its *local-view*, by merging *viewSubset_{v1}* with its current *random-view* in a buffer, selecting R_{size} entries randomly, and updates its *random-view*.

User *u* performs Semt as described in Algorithm 2 with the modification of Figure 5: It computes the rank of each relevant user *v* in its *semantic-view* and adds them to *RankList*. Then it selects the relevant user v_2 that has the highest rank to gossip with. Then it selects a *viewSubset_u* from the *RankList* that have the highest rank and sends it to v_2. Afterward, *u* receives a *viewSubset_{v2}* from v_2. Once user *u* has received, *viewSubset_{v2}*, it updates its *semantic-view* as follows: 1) It computes the rank of each relevant user *v* in the *viewSubset_{v2}* and adds it to the *RankList*. 2) It computes the rank of each relevant user *v* in its *random-view* and adds to the *RankList*. 3) It selects S_{size} entries from the *RankList* which have the highest rank.

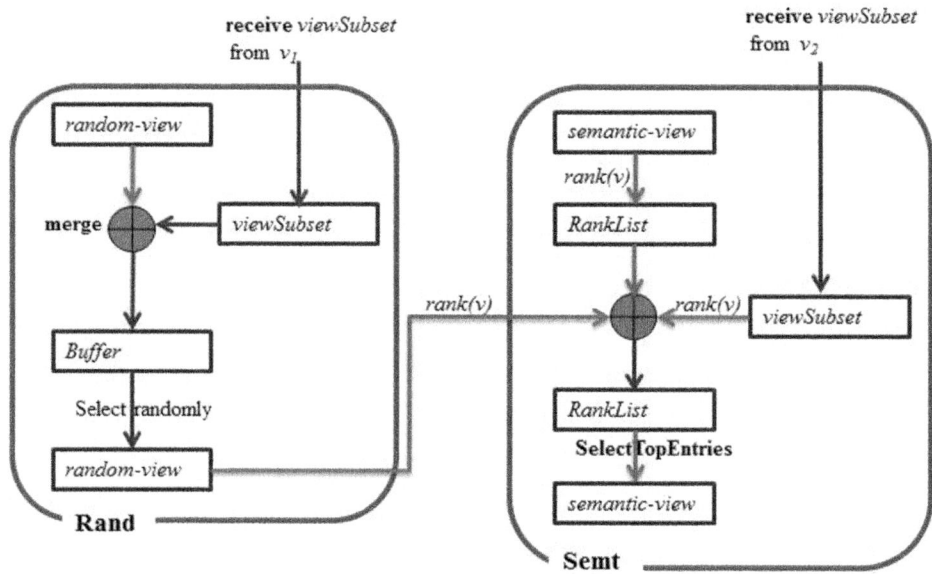

Fig. 7. The 2LG framework at user *u*

7 Query Routing and Result Ranking

In this section, we first describe the query processing algorithm that we use to generate recommendations. Then, we describe the ranking model we use to order the returned recommendations. Finally, we show how users can manage query failures.

7.1 Query Processing

We assume keyword queries of the form $q = \{word_1, word_2, ..., word_l\}$, where l is the number of keywords in the query and $word_i$ is the ith keyword in q. Query q can be of type *push* or *pull* [18]. In the push type, the system automatically extracts the keywords of the query q from the documents that are belonging to the user's topics of interest, such as the most representative words in topics of interest. In the pull type, user u issues a query q with keywords. For both types, the system extracts q's topic vector, denoted by $V_q = [w_q^{t_1}....w_q^{t_d}]$, using LDA as we did for a document. Then query topic(s) $T_q \subset T$ are extracted as described in Section 4.2.

Based on this assumption, each query q issued by a user u has the form $q(word_i,$ TTL, $V_q, T_q, u)$. Algorithm 3 illustrates the behavior of query processing of each user u. In active behavior, u issues a query q and proceeds as follows. First, it selects from its *local-view* the relevant users that are similar to q in terms of topics. Then, it redirects q to those relevant users after reducing the query TTL by one (lines 1 to 6). In other words, user u selects each relevant user $v \in u$'s *local-view* that are similar to q, i.e., *distant(q,v)* $\neq 0$, and then redirects q to them. If 2LG is used, u's *local-view* is the union of its *random-view* and its *semantic-view*.

If user u does not find any relevant user v in its *local-view* that is similar to q, the query q is considered failed, and u uses the *query-history* of the users in its *local-view* to support q (lines 7 to 14) (presented in Section 7.3). Once user u receives the recommendation information from the responders, it ranks those recommendations based on their popularity and semantic similarity (lines 15 to17) (presented in Section 7.2).

In the passive behavior, when user u receives a query q, it processes q as follows. First, u selects from its *local-view* the relevant users that are similar to q, and redirects the query to them if the query TTL is not yet zero (lines 9 to 16). Second, user u measures the similarity between query q and each document user u has locally (lines 3 and 4). The similarity between a document *doc* and q, denoted by *sim(doc,q)*, is measured by using the cosine similarity [32] between the document topic vector $V_{doc}=[w_{doc}^{t_1}....w_{doc}^{t_d}]$ and the query topic vector $V_q = [w_q^{t_1}....w_q^{t_d}]$ which is:

$$sim(doc, q) = \frac{\sum_{i=1}^{d} w_q^{t_i} * w_{doc}^{t_i}}{\sqrt[2]{\sum_{i=1}^{d} w_q^{t_i} * w_q^{t_i} * \sum_{i=1}^{d} w_{doc}^{t_i} * w_{doc}^{t_i}}}$$

Finally, u returns to the query initiator the recommendations for the documents whose similarity exceeds a given (system-defined) threshold (lines 5 and 6).

Algorithm 3- Query Processing
//Active behavior: **Route-Query**(q, local-view_u)
Input: query q (word_i, TTL, V_q, T_q,u); local-view_u
Output: submit q to potential relevant users

```
1    For each relevant user v∈ local-view_u do
2        If distant(q,v) ≠0 then
3            u send q to v
4                q.TTL = q.TTL-1
5        End if
6    End For
7    If query-fail  then
8        For each user v∈ local-view_u do
```

9   user u **retrieve** user v *query-history H_v*
10   **If** $distan(q,q_i) \neq 0$ and $s_{qi}=1$ s.t. $q_j \in H_v$ **then**
11   User u **Send** q to user v
12   **End If**
13   **End For**
14   **End If**
15   **If** user u **Receives** $rec_1,..., rec_n$ **then**
16   User u **Ranks** *($rec_1,..., rec_n$)*
17   **End If**
//Passive behavior: **Process-query**$(q, D_u, local\text{-}view_u)$
Input: query q ($word_i$, TTL, V_q, T_q,u); $local\text{-}view_u$
Output: answer set of information recommendations for query q; u send q to potential relevant users
1   **Forever do**
2   **Receive** query q
3   **For** each $doc \in D_u$ **do**
4        $Sim(q,doc)$ = **CosineSimilarity**(V_q, V_{doc})
5   **If** $Sim(q,doc)$ greater than *threshold* **then**
6   **recommend** *doc* to q's initiator
7   **End If**
8   **End For**
9   **If** $q.TTL$ not equal to zero **then**
10   **For** each relevant user $v \in local\text{-}view_u$ **do**
11   **If** $distant(q,v) \neq 0$ **then**
12   u **send** q to v
13              $TTL = TTL\text{-}1$
14   **End if**
15   **End For**
16   **End If**

With such query routing, we avoid sending q to all neighbors, thus minimizing the number of messages and network traffic for q.

7.2   Ranking Recommendations

Assume the query initiator receives $rec_q^{v1}(doc_1),..., rec_q^{v}(doc_i)$ from the responders, where $rec_q^{v}(doc_i)$ is the recommendation that has been given for a document doc_i from a responder v. $rec_q^{v}(doc_i)$ includes the similarity between query q and document doc_i. With this, the initiator ranks $rec_q^{v1}(doc_1),..., rec_q^{v}(doc_i)$ to provide *recommendations$_q$*. The recommencations $rec_q^{v1}(doc_1),..., rec_q^{v}(doc_i)$ are ranked based on their popularity and semantic similarity (line 16 in the active behavior of Algorithm 3). That is, the rank of a $rec_q^{v}(doc)$, denoted by $rank(rec_q^{v}(doc))$, reflects its semantic relevance with q and its popularity:

$$rank\big(rec_q^v(doc)\big) = a * sim(doc,q) + b * pop(doc)$$

where a and b are scale parameters such that $a + b = 1$ and $pop(doc)$ is the popularity of doc. The popularity is equal to the number of replicas this document has, i.e., the number of users that store a replica of doc. The user can specify whether it prefers highly popular documents or documents that are highly semantically relevant by adjusting parameters a and b. Upon receiving recommendation documents, a user u can download a copy of a document, give a rating to it and include it in its document set D_u.

In the example of Figure 8, suppose that user u initiates a query q for topic t_1 with TTL=2. User u redirects the query q to relevant users v_3 and v_4 after reducing the TTL by 1. When v_3 receives q, it computes the similarity between q and its documents $sim(doc,q)$ where $doc \in D_{v3}$. It then returns to u the recommendations $rec_q^{v3}(doc_i)$ for those documents whose similarity exceeds a given threshold. User v_3 stops redirecting q even though its TTL is not zero. This is because v_3 does not have a relevant user in its *local-view* that is similar to q, and has not yet received a copy of q. Similarly user v_4 computes $sim(doc,q)$ for $doc \in D_{v4}$, and returns the recommendations $rec_q^{v4}(doc_i)$ to u. It then redirects q to relevant user v_1 after reducing TTL by one. From there user v_1 computes $sim(doc,q)$ for $doc \in D_{v1}$ and returns the recommendations $rec_q^{v1}(doc_i)$ to u. User v_1 does not froward q because its TTL has reached zero. Notice that in the case in which u does find any relevant user in its *local-view* that can serve q, it exploits the *query-histories* of the users in its *local-view*.

Once the user u receives $rec_q^v(doc_i)$ from the relevant users v_3, v_4 and v_1, it ranks $rec_q^v(doc_i)$ based on their popularity and semantic similarity to provide recommendations$_q$.

$$rank(rec_q^v(doc)) = a * sim(doc, q) + b * pop(doc)$$

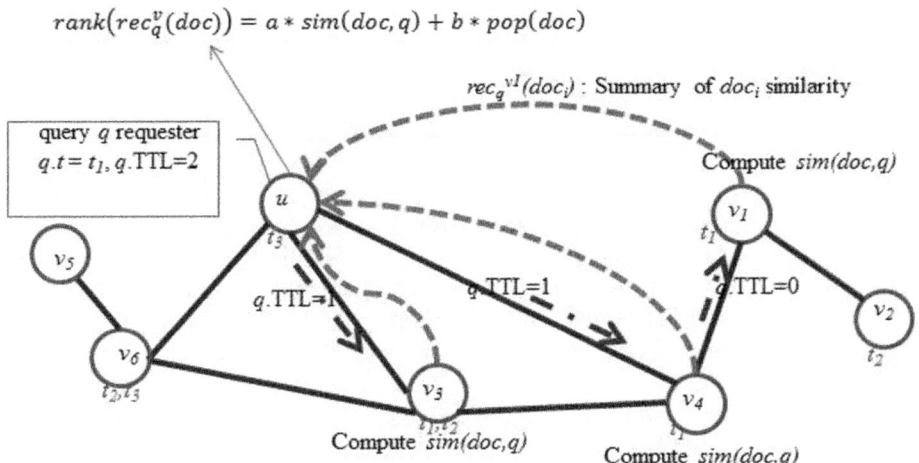

Fig. 8. The query processing and recommendation ranking

7.3 Dealing with Query Failures

We use *query-histories* to support failed queries in the hope to increase the *hit-ratio*. Whenever, a user u submits a query q, it adds q's topics T_q to its *query-history* along with a state, which indicates if q was successfully submitted or not. q was successfully submitted, if u has relevant user(s) in its *local-view* that are similar to q. The idea is that such users can serve other queries that are similar to q.

Query q is considered as *query-fail* if user u does not find any relevant user in its *local-view* that is similar to q, i.e., *distant(q,v)= 0*, for each relevant user $v \in$ *local-view$_u$*. To handle this situation, we exploit the *query-histories* of the users in u's *local-view*.

Recall that each user u maintains a *query-history H_u*. When u experiences a *query-fail*, u retrieves the *query-history H_v* of each user v in its *local-view*. Then, for each H_v, it computes the similarity between q and each query $q_i \in H_v$ (lines 8 and 9 in the active behavior of Algorithm 3). If there is a query q_i such that *distant(q,q_i)≠0* and $s_{qi}=1$, u sends q to v (lines 10 and 11 in the active behavior of Algorithm 3). Notice that we do not use *query-histories* in passive behavior.

8 Experimental Evaluation

In this section, we provide an experimental evaluation of P2Prec to assess the quality of recommendations, search efficiency (cost, and *hit-ratio*), bandwidth consumption, and clustering coefficient. We have conducted a set of experiments using TREC09 [29]. We first describe the experimentation setup. Then, we evaluate each gossip algorithm and its effect on the respective metrics, and the effect of TTL and *query-histories* on query processing.

8.1 Experimentation Setup

We use the classical metric of *recall* in IR and RSs to assess the quality of the returned results [33]. Recall represents the system ability to return all relevant documents to a query from the dataset. Thus, in order to measure recall, the relevant document set for each query that has been issued in the system should be known in advance, i.e., we need to have relevance judgments for each query that has been issued in the system. Data published by TREC have many relevance judgments. We use the Ohsumed documents corpus [15] that has been widely used in IR. It is a set of 348566 references from MEDLINE, the on-line medical information database, consisting of titles or abstracts from 270 medical journals over a five year period (1987-1991). It was used for the TREC09 Filtering Track [29]. It includes a set Q of 4904 queries. The relevant documents for each query q denoted by R_q were determined by TREC09 query assessors. In the experiment, user u issues a query $q \in Q$ and uses P2Prec to possibly retrieve the documents that have been in R_q. The set of documents returned by P2Prec for a user u of a query q is denoted by P_q. Once a user u has received P_q from P2Prec, it can count the number of common documents in both sets P_q and R_q to compute recall. Thus, recall is defined as the percentage of q's relevant documents $doc \in R_q$ occurring in P_q with respect to the overall number of q's relevant documents $|R_q|$:

$$Recall = 100 \cdot \frac{|P_q \cap Rq|}{|Rq|}$$

We use the following metrics to evaluate P2Prec.

— **Communication cost:** the number of messages in the P2P system for a query.
— **Hit-ratio:** the percentage of the number of queries that have been successfully answered.
— **Background traffic:** the average traffic in bps experienced by a user due to gossip exchanges.
— **Clustering coefficient:** the density of connections between peer neighbors. Given a user u, the clustering coefficient of u is the fraction of edges between neighbors (users at u's *local-view*) of u that actually exist compared to the total number of possible edges which is:

$$C.coef_u = \frac{\sum_{i=1}^{view\text{-}size_u} loacl\text{-}view_u \cap loacl\text{-}view_{u_i}}{(view\text{-}size_u)(view\text{-}size_u + 1)}, \quad u_i \in loacl\text{-}view_u$$

In order to compute the clustering coefficient of the network, we sum the clustering coefficient of each user u, and divide it over the number of users in the network.

We extracted the titles and the abstracts of TREC09 documents and removed from them all the stop words (e.g., the, and, she, he, ...) and punctuations. Then, we fed them to the GibbsLDA++ software [24], a C++ implementation of LDA using Gibbs sampling, to estimate the document topic vectors V_{doc}. With $/T/=100$ as the number of topics, we ran GibbsLDA++ 2000 times to estimate the document topic vectors V_{doc}. To estimate the query topic vectors V_q, we removed the stop words and punctuations from queries keywords, fed the query keywords left to the GibbsLDA++, and computed the topics T_q of each query $q \in Q$. We consider that each query $q \in Q$ has one topic $t \in T$ for ease of explanation. We consider as topic t_q of query the maximum component of its V_q, i.e., the maximum w_q^t.

We use a network consisting of 7115 users. Once a user u joins the network, it initializes its local-view by selecting randomly a set of users, and adding them into its local-view (as described in Section 4.3). Suppose that the document popularity follows the zipf distribution [6]. Thus, we assume that the number of replicas of a document is proportional to the number of times a document is relevant for a query in Q. After distributing randomly the TREC09 documents over the users in the network, these users have 693308 documents, with an average of 97.433 documents per user.

We generate a random rating between 0 and 5 for each document a user has and compute the users' topics of interest from the documents they have rated. We consider that each user u is interested at least in one topic, and relevant at least for one topic. Also u is interested at maximum in 10 topics, and relevant at maximum for 5 topics.

P2Prec is built on top of a P2P content sharing system which we generated as an underlying network of 7115 peers (corresponding to users). We use PeerSim [23] for simulation. Each experiment is run for 24 hours, which are mapped to simulation time units.

In order to evaluate the quality of recommendations, we let each user u issue a query after receiving the results from all the users that have received the previous query or after the query has reached a system-specified timeout. The query topic is selected, using zipf distribution, among u's topics of interest T_u. Then, we obtain the recommendations for each query and compute recall, communication cost, and response time. In order to obtain global metrics, we average the respective metric values for all evaluated queries.

Table 1. Simulation Parameters

Parameter	Values
Topics (T)	100
TTL	1, 2, 3
Local-view size (*view-size*)	70
Gossip length (L_{gossip})	20
Gossip period (C_{gossip})	30 min
Random-view size (R_{size})	40
Semantic-view size (S_{size})	30
Gossip period for random at 2LG (C_{random})	10 min
Gossip period for semantic at 2LG ($C_{semantic}$)	30 min

Table 1 summarizes the simulation parameters that we have used in the experiments. We do not study the effect of gossip parameters (*view-size*, L_{gossip} and C_{gossip}) on the recommendation quality (see [10] for such study).

8.2 Experiments

We performed our experiments under churn, i.e., the network size continuously changes during the run due to users joining or leaving the system. The experiments start with a stable overlay with 355 users. Then, as experiments are run, new users are joining and some of the existing users are leaving.

We investigate the effect of Rand, Semt and 2LG on the quality of recommendations over the respective metrics. In each experiment, we run one gossip algorithm (Rand, Semt or 2LG). Then, we collect the results for each algorithm after 24 simulation hours. We set TTL to 1 to measure the quality and effectiveness of users' views.

Table 2 shows the results obtained from the experiments. In [10], we showed that the background traffic is affected by gossip period (C_{gossip}) and gossip length (L_{gossip}). We observe that increasing either C_{gossip} or L_{gossip} increases background traffic while decreasing either C_{gossip} or L_{gossip} decreases it. Thus, Rand and Semt are used with the same gossip parameters (C_{gossip} = 30 minutes, and L_{gossip} = 20), so they consume almost the same bandwidth (4 bps). 2LG consumes more bandwidth because four exchange messages are applied each 30 minutes (three exchanges for Rand and one exchange for Semt). Thus, the background traffic in 2LG is four times that of Rand and Semt (13.979 bps).

Rand produces an overlay with a low clustering coefficient. There is a low overlap between a user u's *local-view* and the *local-views* of its neighbors (the users at u's *local-view*). Semt produces a high clustering coefficient. There is a high overlap between users' *local-views*. This is due to the fact that, if a user u_1 is similar to user

u_2, and user u_2 is similar to u_3, then most probably u_1 and u_3 are similar, and thus produce a clique. In 2LG, the clustering coefficient is moderate between it uses both Rand and Semt,the first favoring randomness while the other favors cliques. Therefore, the clustering coefficient is higher than in Rand but lower than in Semt.

Table 2. Results

Metric	Random	Semt	2LG
Recall	30.7	48.71	42.23
Communication cost	5.04	17.56	10.89
Max. Hit-ratio	0.515	0.835	0.795
Background traffic (bps)	3.484	3.976	13.979
Clustering coefficient	0.073	0.358	0.133

In Figure 9, we show the variation of recall, communication cost, and *hit-ratio* versus time for the three algorithms. Figure 9(a) shows that the recall keeps increasing at the beginning, and almost stabilizes after 10 hours. At the beginning, the network size is small and many relevant users are not alive. Thus, many irrelevant documents are returned, which reduces recall. Semt increases recall by a factor of 1.6 in comparison with Rand and by a factor of 1.12 in comparison with 2LG. This is because in Semt, a user u has in its *local-view* a high number of relevant users that are similar to u's queries. Thus, when u submits a query q, q reaches more relevant users, and thus more relevant documents are returned.

Figure 9(b) shows the communication cost of queries for the three algorithms. We set TTL to 1 so that communication cost represents the number of relevant users that serve the query. We observe that Semt has the highest communication cost, because each user u includes in its *local-view* a high number of relevant users that are similar to u's demands, and thus, each query is sent to many neighbors. In Rand, the communication cost is low because each u has few relevant users to which queries could be sent to. In 2LG, the communication cost is a little less than Semt, because the *semantic-view* size ($S_{size} = 30$) is less than that in Semt (*view-size* = 70).

Figure 9(c) shows the *hit-ratio* for the three algorithms. The maximum *hit-ratio* that has been obtained by Rand is low (0.515). Under Rand, each user u has few relevant users that are similar to u's queries. Thus, when u submits a query q, there is a high probability that u does not find a relevant user in its *local-view* that can serve q. In Semt and 2LG, the *hit-ratio* is high because u's *local-view* includes many relevant users that are similar to u's demands. Thus, when u submits a query q, u finds many relevant users in its *local-view* that can serve its query q.

In Figure 9, the significant jump in the beginning of the results is because we start from time zero, no queries are issued and thus no result is gathered. We start the experiments with a small network size, so the number of involved users is small. Therefore, the average value of the metrics starts large, because we divide the gathered value of metrics over a small number of users. As the experiments proceed and more users join the networks, the number of users involved increases and the average value of the metrics stabilizes.

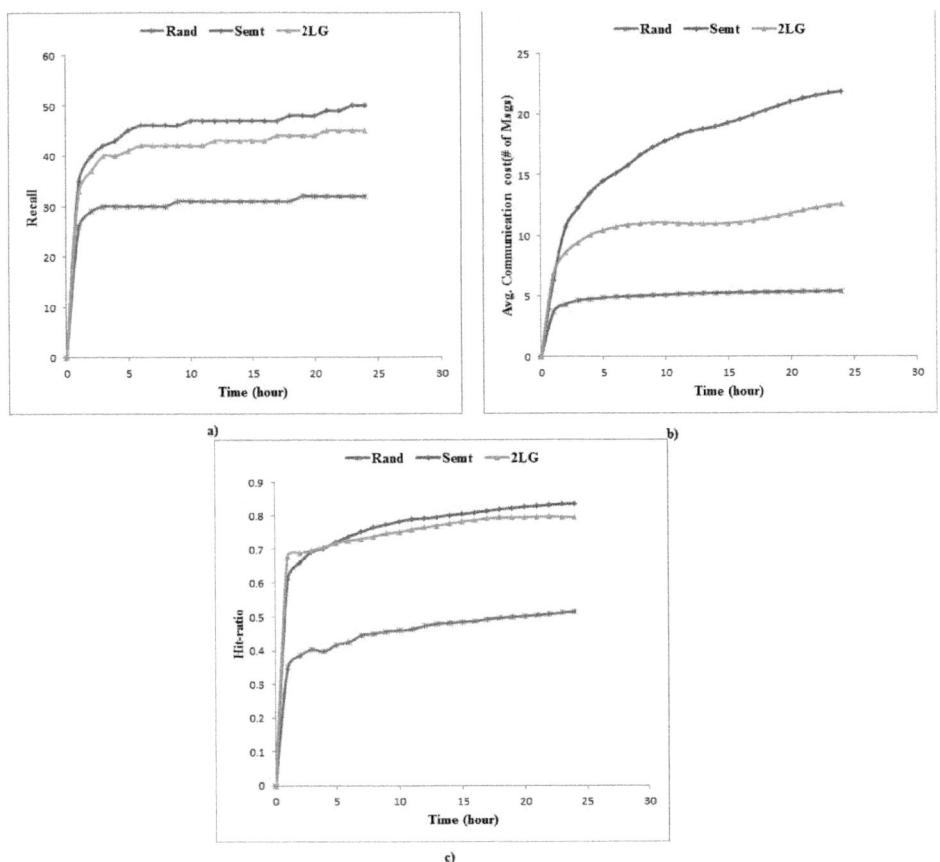

Fig. 9. The variation of recall, communication cost, and *hit-ratio* versus time

8.2.1 Effect of TTL

We investigate the effect of varying TTL on the quality of recommendations over the respective metrics. In each experiment, we run one gossip algorithm and vary TTL. Then, we collect the results for each algorithm under each TTL after 24 simulation hours. We do not show all the results but we explain our observations.

The TTL variation has significant impact on recall and communication cost especially when Rand is used. When increasing TTL, more relevant users are visited, thus increasing the communication cost and the number of returned documents, which in turn increases recall. In Rand, the communication cost is multiplied by 26.5 when TTL increases from 1 to 3 (141 relevant users are visited), while recall is increased from 30% to 73.04%. In Semt, recall is increased from 48% to 68.5%, when TTL increases from 1 to 3, while communication cost is multiplied by 4.62 (100 relevant users are visited). Varying TTL does not have significant impact on Semt, due to the fact that the users' *local-views* have high overlap. Thus, when a user u submits a query q to a user v, v does not have many relevant users in its *local-view* that have not received q before, because the overlap between u's *local-view* and v's *local-view* is high. However, the TTL variation has moderate impact on 2LG. Hence, recall is increased from 42% to 82.4% when TTL increases from 1 to 3, while communication cost is multiplied by 19 (234 relevant users are visited). Remember that in 2LG, each user u uses its random and semantic view. Thus, when a user u submits its query q to a user v, v may find many relevant users in its semantic and random views that have not received q yet.

8.3   Effect of Using Query-histories

In this experiment, we study the effect of using *query-histories* to support failed queries. Figure 10 shows the effect of using *query-histories* in the 2LG algorithm with TTL=1 on recall, communication cost and *hit-ratio*. In the fact, the use of *query-histories* increases the *hit-ratio* to 96.9%. That is, each time a user u submits a query q, there is a high probability to find a relevant user to serve its query either from its view or from its neighbors' *query-histories*. Recall that each user u maintains in its *semantic-view* the relevant users that are most similar to itself. The queries that have been requested by user u are most probably similar to queries that are requested by the users in its *semantic-view*. Thus, when u uses the *query-histories* of the users in its views, it most probably finds a user v that can serve its query.

Using *query-histories* increases recall slightly, because more users are visited and thus, more documents are returned. But it also increases communication cost, because more relevant users are visited.

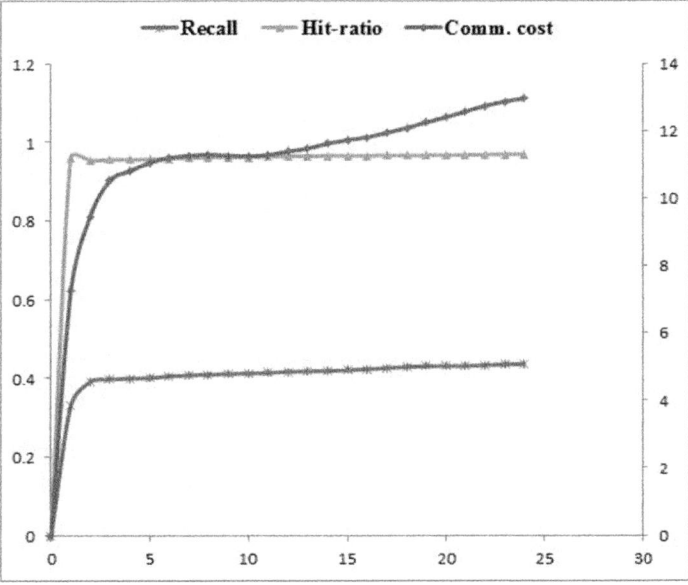

Fig. 10. The effect of *query-histories* on recall, communication cost and *hit-ratio*

9   Related Work

P2Prec leverages collaborative filtering and content based recommendation. It also uses gossip algorithms to manage relevant users for recommendations. Work most related to ours includes centralized recommender systems (RS), distributed RSs, social P2P networks and P2P routing protocols.

Centralized RSs. RSs have been used in major applications such as e-commerce systems like Amazon.com. RSs can be based on collaborative filtering (CF) [13] or content-based recommendations [4]. CF can recommend products to a user based on the products previously rated by similar users. It works by measuring the similarity between users based on the ratings they have given [13]. However, CF suffers from data sparsity as users typically only rate a small numbers of products, making it hard to find similar users. Furthermore, a new user that has not yet rated any product will not find similar users to help in finding recommendations [1].

Content-based filtering has been introduced to alleviate the sparsity problem of CF [4]. Content-based RSs work by suggesting products that are similar to those that the user has seen or rated [4]. The similarity measure is computed between the products the user has seen and the nominated products. Products with high similarity are suggested to the user. However, a user is limited to receive products that are only similar to those it has rated, and thus, might not explore new interesting topics. In P2Prec, each user maintains a list of relevant users with different topics of interest so a user can search a variety of topics even though it is not interested in those topics.

Distributed RSs. Recently, there has been some work on distributed RSs. The first work is Tveit [37], a P2P RS to suggest recommendations for mobile customers. The system is based on a pure P2P topology like Gnutella and queries that include users' ratings are simply propagated by flooding. Miller et al [22] explore P2P RS and propose four architectures including random discovery (similar to Gnutella), transitive traversal, distributed hash table (DHT), and secure blackboard. In these architectures, a user aggregates the rating data from the system to make recommendations. However, aggregating users' ratings to make recommendations increases the amount of traffic within the network. In contrast, P2Prec lets each user maintain locally a set of relevant users along with their topics of interest, so the user uses them to support its queries.

Kermarrec et al. [20] use gossiping and random walk for a decentralized RS. Gossip algorithms let each user aggregate the most similar users (*neighbors*). Information about them is stored locally along with their profiles. A user's profile includes the items the user has rated. Each user computes its recommendations by running locally a random walk on its neighbors and neighbors of neighbors. However, computing the transition similarity matrix of the random walk and its inverse is complex, and is not scalable as the number of items at the user and its neighbors (and neighbors of neighbors) is increased.

Social P2P networks. These systems consider a user's social behaviors in constructing the overlay and searching content. Pouwelse et al. [26] build a distributed system that lets users record videos based on recommendations from other users that have similar habits. The system uses a small number of users to get relevant recommendations from users' profiles. Fast et al. [11] propose knowledge discovery techniques for overlay network creation by clustering users based on their preferences on music style. However, a user's preference locality is used in a static way. Upadrashta et al. [38] propose to enhance the search in Gnutella using social network aspects. In their model, each peer maintains n peers that have strong relationships for each category of interest, and use it for searching files in the network. However, they do not have any mechanism to keep the profile size manageable.

Gossip algorithms were initially proposed for network monitoring. With the challenges brought by distributed and large-scale data management in different domains, gossiping has been used for other purposes. For instance, in [2, 40, 39], the authors propose to introduce semantic parameters while gossiping to design a personalized P2P system. Each user maintains a set of users (*neighbors*) with similar interest along with their profiles. Two layers of gossip are used to find and construct the users' *neighbors*. The first layer is used to aggregate the top similar users. The second layer provides the first layer with new users. Two users are considered similar if they share a common number of items. When a user issues a query, it uses the profiles of its *neighbors* to process locally its query.

These systems have several problems. First, having users maintaining locally the profiles of their *neighbors* leads to storage and inconsistency problems. In P2Prec, each user maintains only its documents, thus eliminating this problem. Second, users construct their *neighbors* by gossiping their profiles, which may yield high network bandwidth consumption. In contrast, P2Prec uses gossip only to exchange the topics of interest between users and the topics of interest are small. Finally, these systems measure the similarity between users based on the number of common items in their profiles. Unfortunately, this kind of similarity does not capture the context of items. For instance, suppose that a user u is interested in topic "*computer science*" and has a set of documents which are related to that topic. Also suppose that another user v is interested in "*computer science*" and maintains another set of documents in that topic, but with no document in common. These users will be considered dissimilar, and thus, will not be related in the overlay. In P2Prec, we would capture this similarity, which is based on their topics of interest.

P2P Routing algorithms. There is a large body of work on search methods in P2P systems (see [25] for a survey). P2P routing algorithms are classified as structured and unstructured. Typically they differ on the constraints imposed on how users are organized and where shared contents are placed [27]. Structured P2P systems such as (Oceanstore [21], CAN [28], Pastry [30], CHORD [35], and Tapestry [41]) impose specific constraints on the network structure. They use a DHT to control content placement and location, and thus, provide an efficient, deterministic search that can locate content in a small number of messages. PSearch [36] uses semantic indexing to distribute documents in the CAN DHT such that documents which are relevant to a query are likely to be collocated on a small number of peers. Sahin et al. [31] represent each document as a vector of terms and uses them to determine the location of the document indices in a Chord DHT. Although structured P2P systems are good for point queries, they are not efficient in locating text and key-words queries.

Unstructured P2P networks impose few constraints on users' neighborhood and content placement [27] so that users in the overlay get loosely connected. They use blind flooding [27] algorithms to disseminate discovery messages or queries. With blind flooding, a user sends a query to all its neighbors. Recently, relationships between content or users and semantic indexing have been proposed to eliminate blind flooding in unstructured P2P networks. For instance, the authors of [3, 34, 8] use information retrieval (IR) techniques to group the peers with similar contents into the same clusters (i.e., peers with similar contents are grouped in one cluster). A peer p is considered belonging to a cluster y, if p has a significant number of contents that are similar to the contents that have been stored by the peers in cluster y. A query is guided to a cluster that is more likely to have answers to the given query and the query is flooded within this cluster. But flooding consumes a lot of network bandwidth, and these solutions do not exploit users' topics of interest as we do in P2Prec.

10 Conclusion

In this paper, we proposed P2Prec, a recommendation system for large-scale data sharing that leverages collaborative filtering and content based recommendation. P2Prec is useful to recommend to a user documents that are highly related to a specific topic from relevant users in that topic. Each user in the system is automatically assigned topics of interest, based on a combination of topic extraction from its documents (the documents it shares) and ratings. To extract and classify the hidden topics available in the documents, we use the LDA technique. P2Prec is built on top of an unstructured overlay for the sake of scalability and decentralization, and uses gossip algorithms to disseminate relevant users and their topics. P2Prec uses two new semantic-based gossip algorithms (Semt and 2LG) to let each user aggregate similar relevant users and insure randomness in its view.

The next step of this work is to design a decentralized RS based on explicit personalization (friendship network) and users' topics of interests over a distributed graph. In addition, we plan to design an algorithm to compute the popularity of a recommended document.

In our experimental evaluation, using the TREC09 dataset, we showed that using Semt increases recall and *hit-ratio*. This is because each user maintains in its *local-view* a high number of relevant users that can serve its demands. Using Rand decreases the overlap between users' *local-views* and thus, increases randomness. Using 2LG exploits the advantages of Rand and Semt. It increases recall and *hit-ratio* by a factor of 1.4 and 1.6, respectively, compared with Rand, and reduces the overlap between users' *local-views* by a factor of 2.è compared with Semt.

Using gossip style communication to exchange the topics of interest (especially in Semt) increases the system's ability to yield acceptable recall with low overhead in terms of bandwidth consumption. Furthermore, it increases the *hit-ratio* because gossiping brings allows similar relevant users to be included into a user's *local-view*, thus reducing the possibility that the user does not find relevant users satisfying its queries.

References

1. Adomavicius, G., Tuzhilin, A.: Toward the next generation of recommender systems: a survey of the state-of-theart and possible extensions. IEEE TKDE 17(6), 734–749 (2005)
2. Bai, X., Bertier, M., Guerraoui, R., Kermarrec, A.M., Leroy, L.: Gossiping personalized queries. In: EDBT, pp. 87–98 (2010)
3. Bawa, M., Manku, G.S., Raghavan, P.: SETS: Search enhanced by topic segmentation. In: ACM SIGIR, pp. 306–313 (2003)
4. Billsus, D., Pazzani, M.J.: Learning collaborative information filters. In: ICML, pp. 46–54 (1998)
5. Blei, D.M., Ng, A.Y., Jordan, M.I.: Latent Dirichlet Allocation. JMLR 3, 993–1022 (2003)
6. Breslau, L., Cao, P., Fan, L., Phillips, G., Shenker, S.: Web Caching and Zipf-like Distributions: Evidence and Implications. In: INFOCOM, pp. 126–134 (1999)
7. Callan, J.: Distributed Information Retrieval. In: Croft, W.B. (ed.) Advances in Information Retrieval, pp. 127–150. Kluwer Academic Publishers, Dordrecht (2000)
8. Crespo, A., Garcia-Molina, H.: Semantic Overlay Networks for P2P systems. Technical report, Stanford University (2003)
9. Dice, L.R.: Measures of the Amount of Ecologic Association between Species. Ecology 26(3), 297–302 (1945)
10. Draidi, F., Pacitti, E., Valduriez, P., Kemme, B.: P2Prec: a Recommendation Service for P2P Content Sharing Systems. In: BDA (2010)
11. Fast, A., Jensen, D., Levine, B.N.: Creating social networks to improve peer to peer networking. In: ACM SIGKDD, pp. 568–573 (2005)
12. Gavidia, D., Voulgaris, S., Steen, M.: Cyclon: Inexpensive Membership Management for Unstructured P2P Overlays. JNSM 13(2), 197–217 (2005)
13. Goldberg, D., Nichols, D., Oki, B., Terry, D.: Using Collaborative Filtering to Weave an Information Tapestry. Commun. ACM 35(12), 61–70 (1992)
14. Hauser, C., Irish, W., Larson, J., Shenker, S., Sturgis, H., Swinehart, D., Demers, A., Greene, D., Terry, D.: Epidemic Algorithms for Replicated Database Maintenance. In: ACM PODC, pp. 1–12 (1987)
15. Hersh, W.R., Buckley, C., Leone, T., Hickam, D.H.: Ohsumed: An interactive retrieval evaluation and new large test collection for research. In: ACM SIGIR, pp. 192–201 (1994)
16. Jelasity, M., Montresor, A.: Epidemic-style Proactive Aggregation in Large Overlay Networks. In: ICDCS, pp. 102–109 (2004)
17. Jelasity, M., Voulgaris, S., Guerraoui, R., Kermarrec, A.M., VanSteen, M.: Gossip-based peer sampling. ACM TOCS 25(3) (2007)
18. Kendall, J., Kendall, K.: Information delivery systems: an exploration of web pull and push technologies. Commun. AIS 1(4), 1–43 (1999)
19. Kermarrec, A.M., Eugster, P.T., Guerraoui, R., Massoulieacute, L.: Epidemic Information Dissemination in Distributed Systems. IEEE Computer 37(5), 60–67 (2004)
20. Kermarrec, A.M., Leroy, V., Moin, A., Thraves, C.: Application of Random Walks to Decentralized Recommender Systems. In: OPODIS, pp. 48–63 (2010)
21. Kubiatowicz, J., Bindel, D., Chen, Y., Czerwinski, S., Eaton, P., Geels, D., Gummadi, R., Rhea, S., Weatherspoon, H., Weimer, W., Wells, C., Zhao, B.: Oceanstore: An architecture for global-scale persistent storage. In: ASPLOS, pp. 190–201 (2000)
22. Miller, B.N., Konstan, J.A., Riedl, J.: PocketLens, Toward a Personal Recommender System. ACM TOIS 22(3), 437–476 (2004)
23. Peersim p2p simulator, http://www.peersim.sourceforge.net

24. Phan, X.-H., `http://gibbslda.sourceforge.net`
25. Pisson, J., Moors, T.: Survey of research towards robust peer-to-peer networks: search methods. Technical report, Univeristy of New South Wales (2004)
26. Pouwelse, J., Slobbe, M., Wang, J., Reinders, M.J.T., Sip, H., P2Pbased, P.V.R.: Recommendation using Friends, Taste Buddies and Superpeers. In: IUI (2005)
27. Qiao, Y., Bustamante, F.E.: Structured and unstructured overlays under the microscope: a measurement-based view of two P2P systems that people use. In: USENIXATEC, pp. 341–355 (2006)
28. Ratnasamy, S., Francis, P., Handley, M., Karp, R., Shenker, S.: A scalable content-addressable network. In: ACM SIGCOMM, pp. 161–172 (2001)
29. Robertson, S., Hull, D.A.: The TREC-9 filtering track final report. TREC-9, 25-40 (2001)
30. Rowstron, A., Druschel, P.: Pastry: Scalable, decentralized object location, and routing for large-scale peer-to-peer systems. In: Liu, H. (ed.) Middleware 2001. LNCS, vol. 2218, pp. 329–350. Springer, Heidelberg (2001)
31. Sahin, O.D., Emekci, F., Agrawal, D.P., El Abbadi, A.: Content-based similarity search over peer-to-peer systems. In: Ng, W.S., Ooi, B.-C., Ouksel, A.M., Sartori, C. (eds.) DBISP2P 2004. LNCS, vol. 3367, pp. 61–78. Springer, Heidelberg (2005)
32. Salton, G.: A Theory of Indexing. In: Conf. Series in Appl. Math., Soc. For Indust. And Appl. Math., J. W. Arrowsmith Ltd. (1975)
33. Sarwar, B., Karypis, G., Konstan, J., Riedl, J.: Analysis of Recommendation Algorithms for e-commerce. In: ACM COEC, pp. 158–167 (2000)
34. Sripanidkulchai, K., Maggs, B.M., Zhang, H.: Efficient content location using interest-based locality in peer-to-peer systems. INFOCOM 3, 2166–2176 (2003)
35. Stoica, I., Morris, R., Karger, D., Kaashoek, M.F., Balakrishnan, B.: Chord: A scalable peer-to-peer lookup service for internet applications. In: ACM SIGCOMM, pp. 149-160 (2001)
36. Tang, C., Xu, Z., Dwarkadas, S.: Peer-to-Peer Information Retrieval Using Self-Organizing Semantic Overlay Networks. In: ACM SIGCOMM, pp. 175–186 (2003)
37. Tveit, A.: Peer-to-Peer Based Recommendations for Mobile Commerce. In: WMC, pp. 26–29 (2001)
38. Upadrashta, Y., Vassileva, J., Grassmann, W.: Social Networks in Peer-to-Peer Systems. In: HICSS (2005)
39. Voulgaris, S., Van Steen, M.: Epidemic-style management of semantic overlays for content based searching. Technical Report, Amsterdam (2004)
40. Wang, J., Pouwelse, J.A., Fokker, J.E., de Vries, A.P., Reinders, M.J.T.: Personalization of peer-to-peer television system. Or EuroITV, 147–155, 2006
41. Zhao, B., Kubiatowicz, J., Joseph, A.: Tapestry: An infrastructure for fault-tolerant wide-area location and routing. Technical Report, U. C. Berkeley (2001)

Energy-Aware Data Processing Techniques for Wireless Sensor Networks: A Review

Suan Khai Chong[1], Mohamed Medhat Gaber[2],
Shonali Krishnaswamy[1], and Seng Wai Loke[3]

[1] Centre for Distributed Systems and Software Engineering,
Monash University, 900 Dandenong Rd, Caulfield East, Victoria 3145
Australia
[2] School of Computing
University of Portsmouth
Portsmouth, Hampshire, England, PO1 3HE
UK
[3] Department of Computer Science and Computer Engineering
La Trobe University
Victoria 3086
Australia

Abstract. Extensive data generated by peers of nodes in wireless sensor networks (WSNs) needs to be analysed and processed in order to extract information that is meaningful to the user. Data processing techniques that achieve this goal on sensor nodes are required to operate while meeting resource constraints such as memory and power to prolong a sensor network's lifetime. This survey serves to provide a comprehensive examination of such techniques, enabling developers of WSN applications to select and implement data processing techniques that perform efficiently for their intended WSN application. It presents a general analysis of the issue of energy conservation in sensor networks and an up-to-date classification and evaluation of data processing techniques that have factored in energy constraints of sensors.

1 Introduction

Data processing engines in wireless sensor networks (WSNs) are typical energy and processing power constrained P2P systems. There are several challenges that need to be addressed in order to promote the wider adoption and application of WSNs. These challenges relate to both individual sensor hardware and operations of the sensor network as a whole. Individual sensors have limitations in terms of sensing and processing capabilities [48] while on the level of the whole network, the issues extend to finding a suitable communication protocol to deal with frequent topology changes, routing protocols to maximise sensor lifetime and dealing with extensive data generated from sensor nodes [1]. In order to deal with the extensive data generated from sensor nodes, the main strategies that have been proposed in the domain of WSNs involve:

A. Hameurlain, J. Küng, and R. Wagner (Eds.): TLDKS III , LNCS 6790, pp. 117–137, 2011.

- **How frequently to sense and transmit data?** the frequency of transmission is important because radio communication consumes the most energy in a sensor network [2]. Energy consumption is significantly reduced when sensors exchange only data that is necessary for the application, for example, sending data on user demand only.
- **How much data has to be processed?** data exchanged between sensor nodes can be either raw (i.e. sensed readings) or processed (e.g. averaged sensed readings). Processing sensor data enables essential information to be filtered out from all data collected on sensors and communication of only data that is important for the application.
- **How is the data to be communicated?** this refers to the communication and routing protocols to transport sensor data from one sensor to another sensor or base station in the network. For instance, the decision to communicate the data from a sensor directly to the base station or via neighbouring sensors to base station. Communication of data from a sensor node directly to a base station may drain the node's energy significantly due to transmission over a long distance whereas routing via peer nodes may prolong the node's energy but decrease overall network lifetime.

Generally, approaches that deal with decisions about the amount of data to be processed or communicated can be classified as data processing approaches. Approaches that deal with the underlying mechanism to communicate the data are referred to as communication protocols for wireless sensor networks. These two types of approaches can work together to conserve energy. We focus on data processing approaches that efficiently reduce the amount of sensor data exchanged, and thereby prolong sensor network lifetime. These can be divided into approaches that operate at the network level or at the node level, as explained in section 2. Following this, section 2 then further discusses the techniques that work at the network level, while section 3 elaborates on the techniques that work at the node level. Lastly, section 4 draws some conclusions about these techniques.

2   Network-Based Data Processing Approaches

As sensor data communication operation is significantly more costly in terms of energy use than sensor data computation, it is logical to process sensor data at or close to the source to reduce the amount of data to be transmitted [48]. This data originates from the sensing capabilities of sensor nodes and can be either stored to be processed at a later time or treated as a continuous stream to be processed immediately [14].

In this section, we introduce the approaches that process such sensor data with a focus on reducing overall energy consumption. These approaches can be broadly classified to be either: (1) network-based, which refers to approaches that involve processing sensor data at the base station; or (2) node-based, which refers to approaches that involve processing sensor data locally at sensor nodes. The former category, network-based, is discussed in details in this section.

Figure 1 illustrates the data processing at network and node levels. The taxonomy of network and node-based data processing approaches is presented in figure 2.

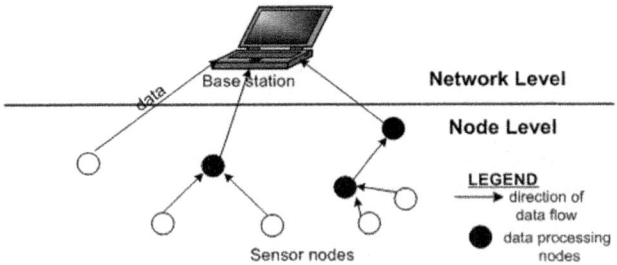

Fig. 1. Processing at network and node levels

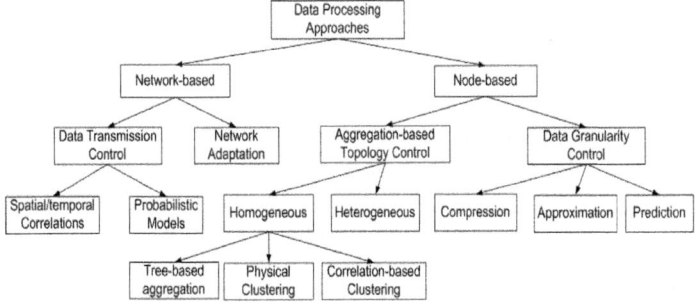

Fig. 2. Data processing approaches taxonomy

Network-based approaches focus on processing of sensory data collected at a resource-rich central location (for instance, a base station) in a sensor network in order to conserve energy. This rationale has been adopted in sensor data sampling techniques, WSN monitoring applications and sensor data querying systems with respect to *data transmission control* and *network adaptation*.

2.1 Data Transmission Control

In this category of approaches, the main idea is that if sensor data correlations or probabilistic models about sensor data is derived at the base station, the base station selectively choose nodes in the sensor network to send their data samples during data gathering. As a consequence, while sampling sensor nodes, the energy consumption of nodes is minimised through reducing the number of data samples that needs to be collected at the base station. In the following section, we discuss existing works that have predicted sensor correlations and sensory values at the base station as a means to efficiently control data collection in this manner.

2.1.1   Spatial/Temporal Correlations

Spatial and temporal correlations have been utilised to more efficiently perform sampling. In [46], the blue noise sensor selection algorithm is presented to selectively activate only a subset of sensors in densely populated sensor networks while sampling. This selection of nodes is derived through building a statistical model of the sensor network nodes known as blue noise pattern (typically used in image processing applications). This model is used to deselect sensor nodes that are active whilst maintaining sufficient accuracy in sensing. The algorithm used also ensures that sensor nodes with the least residual energy are less likely to be selected through incorporating of an energy cost function in the sensor selection algorithm. This approach is an improvement over existing coverage-preserving approaches such as PEAS [60] and node scheduling [55] that do not consider load balancing in node selection.

Another study that utilises spatial correlation to improve sampling is [57] that reduces the number of nodes reporting data to the base station (also referred to as sink). At the sink, an algorithm (known as Iterative Node Selection (INS)) is utilised to derive the number and location of representative nodes in the WSN. These values are derived by using the spatial correlations between sensor observations to choose the representative nodes in the event area whilst, minimising the resulting distortion in sensor data received at the sink. In these approaches, energy is conserved as only a subset of nodes are selected to send data while other sensor nodes can be switched off. A similar idea has also been explored in [59] whereby the authors propose a technique termed 'backcasting' to periodically select only a small subset of sensors with high spatial densities to communicate their information to the base station, whilst ensuring that there is minimal accuracy loss in the data collected.

2.1.2   Probabilistic Models

Probabilistic models have also been used as a means to reduce the amount of communication from sensor nodes to the network's base station [11,56]. In [11], this is achieved through utilising a probabilistic model based on time-varying multivariate Gaussians to derive the probability density function over the set of sensor attributes present in a stored collection of data. The resulting probabilities can then be used to derive spatial and temporal correlations between sensing attributes, which are consequently utilised to form the answer to any user query. Their approach conserves energy as these correlations can be used to develop a query plan that chooses the most cost effective plan to query the sensors. For instance, in answering a query, the model might suggest a plan to sample a voltage sensor (voltage reading used to predict temperature reading) rather than the temperature sensor as the cost to sample a temperature sensor is higher.

In [12], the authors propose a database abstraction known as model-based views to presents users with a consistent view of sensor data queried irrespective of the time and space. This is because the proposed system, MauveDB, applies a predictive model to the static data as well as keeping the output of the model used consistent with changes to the underlying raw data. This allows a conceptual view of sensor data to be presented to the user on demand. The same notion of providing a consistent view of data has also been presented in [28] where a dynamic probabilistic model (DPM) is exploited to create a probabilistic database view.

In another study, [8] propose to conserve energy by efficiently maintaining probabilistic models without sacrifice to sensor communication cost and data quality. More specifically, in their approach, two dynamic probabilistic models are maintained over sensor network attributes: one distributed across and the other at the base station. By keeping these two models in sync at all times, the approach allows data to be communicated only when the predicted values at the base station are not within bounds. Consequently, energy is conserved as the spatial correlations calculated at sensors node can be used to reduce the amount of data communicated from nodes to the base station when answering a query. Similarly, in [56], their approach involves enforcing the building of autoregressive models locally on the node to predict local readings and collectively at the base station. In this case, time series forecasting is employed for prediction.

Separately, in [26], the authors propose an architecture that focus on predicting data reliably at the base station mainly without sensor involvement to minimise communication overhead. This has been achieved through the use of Kalman Filter to filter arriving sensor streams and predict future sensory values. The novelty of the Kalman Filter in their approach lies its ability to predict effectively the internal state of the data stream system in a wide range of wireless sensor applications, for instance in object tracking or network monitoring.

2.2   Network Adaptation

Alternatively, a network-based approach can conserve energy through adapting sensor network operations to physical parameters in the network such as energy levels or to adapt sensing to specific requirements of the WSN monitoring application (for instance, perform sensing only when an expected event is likely to occur). Approaches such as [42,49] have explored these aspects for energy conservation.

In [44], a system level dynamic power management scheme is used, that adapts sensing and communication operations to the expected occurrence (in terms of probability measure) of the event being monitored. This approach is proposed as traditional power management schemes only focus on hardware idleness identification to reduce energy consumption. Their results have shown a gain of 266.92% in energy savings compared to the naive approach that does not apply such adaptation behaviour.

Separately, in [9], the authors propose the use of low-power sensors as tools for enhancing the operating-system-based energy management. From readings obtained by low-powered sensors, the system would infer the user intent and use it to match user needs to conserve energy. For example, energy-consuming cameras that capture user faces would turn off power until low-power thermal sensors indicate a user's presence. They have evaluated this approach by an experimental setting consisting of a camera that periodically captures images and a face detection algorithm that determines the presence or absence of a user. When a user is present at the screen, the display on the computer will be turned on and vice-versa. An average energy saving of 12% has been reported respectively for their prototype using low-power sensors.

In [42], sensor energy is conserved through the use of *energy maps*. The energy maps provide global information about the energy status of all sensors. This information enables an algorithm to make intelligent energy-saving decisions such as selecting packet routes that preserves energy of sensors with low energy reserves. The building of energy maps is, however, an energy-consuming task. The method proposed is to allow each node to calculate its energy dissipation rate from its past readings and forwards the information to monitoring nodes, which can then update this information locally by calculations. They have evaluated, using simulations, their approach by comparing the amount of energy saved using their proposed approach to the approach where each node sends periodically to a monitoring node only its available energy.

Also, in [49], the authors describe a power-efficient system architecture that exploits the characteristics of sensor networks. This architecture manipulates access points with more resources within one hop transmission of sensor nodes to coordinate sensor nodes activities. Three operating phases for a set of sensor nodes and their access points are evaluated: (1) the topology learning phase; (2) the topology collection phase; and (3) the scheduling phase. Both topology learning and collection phases aid the access points in determining a complete topology information of cooperating sensor nodes. With this information, in the scheduling phase, the access points will then determine packet-sensing behaviour of connected sensor nodes. The authors compared their scheme to a random access scheme. Their results have an extended sensor network lifetime of 1 to 2 years relative to 10 days if the random access scheme is applied, transmission of packets are optimally scheduled to avoid retransmissions, while in the random access scheme, energy is lost due to the large number of packet retransmissions that can occur in collisions.

Figure 3 lists the approaches, their main feature and limitation as well as some representative techniques for each application category and the energy savings that have been reported.

In this section, we have discussed network-based data processing approaches that target energy conservation in WSNs. The following section describes

Approach Type	Feature	Limitation	Techniques	Energy Saved
Data transmission control	+ Generic energy conservation techniques.	- Assumes correlated data.	Blue noise sensor selection	85% energy saved compared to random node selection.
			Iterative node selection	Up to 50% energy saved in ns-2 simulation.
			Kalman Filter	10% improvement in saving communication resource.
Network adaptation	+ Techniques optimised to application.	- Application-specific.	Sensing user intention	12% of energy saved for system overall.
			System-level DPM	266.92% energy saved compared to using a random scheme.

Fig. 3. Comparison of network-based approaches

approaches that focus on processing the data on sensor nodes to further reduce network data transmissions and extend network lifetime.

3   Node-Based Data Processing Approaches

Node-based approaches are those that focus on data processing on sensor nodes locally. As illustrated in figure 2, this section discusses two main aspects of energy conservation at the node level:

How data is to be communicated? this concerns data-centric routing techniques used to determine the structure of communication in the WSN in an energy-efficient manner. In figure 2, this is illustrated as *aggregation-based topology control*.

How much/frequently data is to be communicated? this refers to data processing techniques that have been proposed to conserve energy at the node level (*data granularity control* in figure 2).

The following sections describe the aforementioned two categories in greater detail.

3.1   Aggregation-Based Topology Control

A wireless sensor network can be either heterogeneous, whereby the network is made up of nodes with different data processing capability and energy capacity; or homogeneous whereby all nodes have equal data processing capability and energy capacity. In the heterogeneous network, nodes with more resources can automatically be used for data processing or as intermediate nodes for aggregation. For instance, in [22], resource-rich Stargate nodes are responsible for performing integer-only Fast Fourier Transforms and machine learning while other mica2 nodes in the system are only used to collect acoustic data samples. In [32], data processing nodes are iPaqs in their hierarchical network comprised of *macro-nodes* (iPaqs) and *micro-nodes* (mica motes).

However, in the homogeneous network, all nodes have equal capabilities and a data communication structure is necessary in order to balance the communication load in the network. This section focuses on existing literature that have studied how data from a sensor network can efficiently enable energy-efficient routing of network sensor data to the base station, otherwise known as data aggregation. As illustrated in figure 2, the approaches that enables data aggregation can be further divided into: (1) tree-based aggregation; (2) physical clustering; and (3) correlation-based clustering.

3.1.1   Tree-Based Aggregation
Tree-based aggregation describes data-centric routing techniques that apply the idea of a communication tree rooted at the base station; in which data packets from leaf nodes to the base station are aggregated progressively along the tree through the intermediate nodes. Additional data processing can be performed

on the intermediate nodes that route packets in the communication tree. Tree-based aggregation includes tree-based schemes [31,25,35] and variations to the tree-based schemes [43,38].

Early tree-based schemes have been initially discussed in [31] including:

1. The center at nearest source (CNS) scheme, in which data is aggregated at the node nearest to the base station.
2. The shortest paths tree (SPT) scheme, in which data is transmitted along the shortest path from the node to the base station and data is aggregated at common intermediate nodes.
3. The greedy incremental tree (GIT) scheme, in which the paths of the aggregation tree are iteratively combined to form more aggregation points (the GIT scheme has been further evaluated in [24]).

A well-known tree-based scheme is Directed Diffusion [25]. It is a data-centric routing protocol that allows an energy-efficient routing path to be constructed between the base station and the sensor node that answers the query. The protocol works in the following way. When a query is sent from the base station to the WSN, the base station propagates *interest* messages to surrounding nodes nearest to it. This interest message describes the type of sensory data relevant to answering the query. The nodes, upon receiving this interest message, perform sensing to collect information to answer the query and rebroadcasts the message to their neighbours. In this process, every node also sets up a data propagation gradient used to route answers back to the base station along the reverse path of the interest. The intermediate nodes involved in this propagation can perform data aggregation or processing. The energy expenditure from this technique comes from the frequency of the gradient setup between the nodes, which typically requires data exchanges between neighbourhood nodes.

The Tiny AGgregation approach (TAG) [35] is another approach that uses a tree-based scheme for aggregation. In TAG, the communication tree is constructed by first having the base station broadcast messages to all sensor nodes in order to organise the nodes with respect to their node and distance from the base station in terms of levels. For instance, any node without an assigned level that hears a broadcast message will set its own level to be the level in the message plus one. The base station broadcasts this message periodically to continually discover nodes that might be added at a later stage to the topology. Any messages from a sensor node are then first sent to its parent at the level above it and this process is repeated until it eventually reaches the base station. The main energy expenditure in TAG is in the requirement of having nodes in constant listening mode to receive broadcast messages from the base station. As a consequence, less running energy may be consumed in comparison to Directed Diffusion. However, Directed Diffusion has the advantage in saving energy in the long-term due to its use of cost-optimised routes to answer queries.

Other variations to the tree-based scheme include Synopsis Diffusion [43] and Tributaries and Delta [38]. In [43], a ring topology is formed when a node sends a query over the sensor network. In particular, as the query is distributed across to the nodes, the network nodes form a set of rings around the querying node (i.e. the base station). Nevertheless, although the underlying communication is broadcast, the technique only requires each node to transmit exactly once, allowing it to generate equal optimal number of messages as tree-based schemes. However, in this technique, a node may receive duplicates of the same packets from other neighbouring nodes, which can affect the aggregation result. Improving over [43] and tree-based approaches, [38] have proposed an algorithm that alternates between using a tree structure for efficiency in low packet loss situations and the ring structure for robustness in cases of high packet loss.

Topology control protocols can also be clustering-based. In the clustering scheme, cluster heads are nominated to directly aggregate data from nodes within their cluster. The following sections describe the two main types of cluster schemes, namely physical clustering and correlation-based clustering.

3.1.2 Physical Clustering

In physical clustering, the clustering is performed on the basis of physical network parameters such as a node's residual energy. Physical clustering protocols discussed in this literature include Hybrid Energy Efficient Distributed Clustering: HEED [63], Low Energy Adaptive Clustering Hierarchy: LEACH [19] and their variants that run on sensor nodes. Firstly, [19] have proposed the LEACH protocol that allows nodes distributed in a sensor network to organise themselves into sets of clusters based on a similarity measure. Among these sets of clusters, sensors would then elect themselves as cluster-heads with a certain probability. The novelty of LEACH lies in the randomised rotation of high-energy cluster-head selection among sensors in its cluster to avoid energy drain on a single sensor. Finally, local data fusion is performed on cluster heads to allow only aggregated information to be transmitted to the source, thereby enhancing network lifetime. LEACH-centralised (LEACH-C) [20] improves over LEACH in terms of data delivery by the use of an overlooking base station to evenly distribute cluster head nodes throughout the sensor network. The base station does this by computing the average node energy, and that allows only nodes over the average node energy to become cluster heads for the current iteration.

Improving upon LEACH and its variant is the distributed clustering algorithm known as HEED (Hybrid Energy-Efficient Distributed Clustering) [63] for sensor networks. The goal of HEED is to identify a set of cluster heads, which can cover the areas that the sensor nodes monitor, on the basis of the residual energy of each node. This is achieved using a probability function to determine the likelihood a node will become a cluster head in order to select the node that attracts more nodes in terms of proximity and which has most residual energy left. HEED, however, has the drawback that additional sensor energy would be depleted in the changing of the cluster heads at each reclustering cycle. To address this limitation, it is essential to prolong the time between changing the cluster heads and running the HEED protocol over longer time intervals. In terms of evaluations, HEED has shown favourable results compared to other techniques such as LEACH, which selects a random cluster head and also has a successful implementation on Berkeley motes.

Adopting similar notions in HEED, several other techniques have tried to improve HEED/LEACH by developing ways to more efficiently select cluster heads. In [27], sensor nodes communicate among themselves through broadcasts messages to form tighter sensor clusters of closer proximity to one another. The sensors stop their broadcasts when the cluster becomes stable. Their technique has been shown to increase the number of sensor nodes alive over LEACH but a shorter time to first node death. Similarly, [61] improves over LEACH by favouring cluster heads with more residual energy and electing them based on local radio communication to balance load among the cluster heads.

Alternatively, [33] present a chain-based protocol, termed PEGASIS (Power-Efficient Gathering in Sensor Information Systems) that improves over LEACH. The main idea in PEGASIS is for nodes to receive from and transmit data to their close neighbours in a chain-like manner, taking turns as intermediate nodes that would transmit directly to the base station. This is done to reduce the number of nodes communicating directly to the base station. The chain can be created by randomly choosing nodes with favourable radio communication strength or created by the base station, which will broadcast the route that forms the chain to all sensor nodes.

3.1.3   Correlation-Based Clustering

More recently, existing work have shown that cluster heads can also be selected in favour of spatial or temporal data correlations between sensor nodes [62]. One such work is Clustered AGgregation (CAG) [62] in which the authors exploit

Approach Type	Features	Limitations
Tree-based aggregation	+ Only local knowledge of topology required. + Duty-cycling can be implemented on top of aggregation technique to save energy. + Robustness in ring topologies.	- Some tree-based protocols not dynamic in presence of node changes, resulting in energy loss from data retransmissions.
Physical clustering	+ Clustering can be on basis of node residual energy, thereby further prolonging overall network lifetime. + Dynamic to topology changes.	- Energy lost when nodes change their cluster memberships.
Correlation-based clustering	+ Clusters created are correlated in data similarity. Useful to answer approximate data queries.	- Does not adapt to node residual energy.

Fig. 4. Topology control techniques summary

spatial and temporal correlations in sensor-data to form clusters with similar node sensory values within a given threshold and that the clusters remain fixed until the sensory value threshold has changed over time. When the threshold values change, the related sensor nodes will then communicate with neighbouring nodes associated with other clusters to change their cluster memberships. CAG allows the user to derive approximate answers to a query. A similar approach can also be observed in [40] whereby spatial correlations are used to group sensor nodes when they have the same behaviour in movement.

Figure 4 illustrates the qualitative differences between the aforementioned aggregation-based topology control mechanisms. We discuss next how energy can be conserved through data processing regardless of the sensor network topology used.

3.2   Data Granularity Control

In this section, we describe existing approaches that reduce the amount of data communicated in-network and thus, prolong sensor network lifetime. The specific type of approach that can be applied to reduce communication is dependent upon the granularity of data required for the WSN application. For instance, a critical-sensing WSN application such as smart home health care systems [29] would require accurate values from sensor nodes at all times to monitor patient health conditions, whereas coarse-granularity data would suffice for certain event detection systems [17,18]. As a consequence, various data granularity control mechanisms have been proposed to cater to the data requirements of WSN applications. As illustrated in figure 2, these techniques include *compression, approximation* and *prediction*. In the following sections, we describe the approaches in each of these categories in detail, in the context of their purpose in energy conservation.

3.2.1   Compression

Data compression at sensor nodes serves to reduce the size of data packets that are to be transmitted through packet encoding at the sensor node and decoding at the base station. It is desirable to apply compression techniques on sensory data when highly accurate sensory data from the sensing application is required. However, existing data compression algorithms such as bzip2 [52] are not feasible to be directly implemented on sensors due to their large program sizes [30]. Furthermore, as discussed in [4], there is a net energy increase when an existing compression technique is applied before transmission. Compression algorithms for sensor networks are required to operate with a small memory footprint and low-complexity due to sensor hardware limitations. The compression schemes that serve to meet these requirements focus on either improving the efficiency of the coding algorithm (light-weight variants of existing compression algorithms) or utilising the data communication topology to reduce the amount of data to be compressed (for instance, performing compression mainly on intermediate nodes in a data aggregation tree).

Studies that have manipulated the data communication structure to more efficiently perform compression include [47,3]. In [47], a scheme known as 'Coding by Ordering' has been proposed to compress sensory data by encoding information according to the arrival sequence of sensory data packets. The compression scheme merges packets routed from nodes to base station into one single large packet up an aggregation tree. The main idea used in compression is to disregard the order in which sensor data packets reach the base station in order to suppress some packets sent from intermediate aggregator nodes to the base station. In effect, this omits the encoding required for the suppressed packets and thus, improves the efficiency to perform compression. This idea to manipulate sensor communication behaviour to improve compression efficiency is also shared in [3], whereby the rationale used is to buffer sensor data for a specified time duration at the aggregator node's memory. This allows the aggregator node to combine data packets and reduce data packet redundancies prior to transmission. The proposed scheme termed 'Pipelined In-Network Compression' by [3] improves data compression efficiency by allowing sensor data packets to share a prefix system with regard to node IDs and timestamp definitions. In this scheme, the data compression efficiency depends on the length of the shared prefix, i.e. the longer the length of the shared prefix, the higher the data compression ratio.

Spatial and temporal correlations in sensor data have also been manipulated to reduce the compression load [7,21,45]. In [7], the approach is to use the base station to determine the correlation coefficients in sensor data and use the derived correlations to vary the level of compression required at individual sensor nodes. The advantage of this approach is that it reduces the communication load to calculate correlations in-network but assumes the availability of the base station to perform the computation. On the contrary, the approach proposed in [21] uses the idea of computing spatial correlations on sensors locally from data packets overheard on the broadcast transmission channel from neighbouring. This approach allows sensor nodes to collaborate on the data encoding process in order to reduce overall data compression ratio on the node, whilst still enabling the data packets to be decoded exactly at the base station. This in effect reduces the amount of transmission required. The benefits of using spatial correlations with compression have been studied more broadly in [45], whereby the authors explored the energy efficiency in the compression of correlated sensor data sources on sensor data given varying levels of correlations.

Apart from the using physical or data parameters to enhance the data compression technique, it is also important for the designed data compression algorithm to be tailored to resource constraints of sensor hardware. This has been studied in [51,39]. In [51], the authors propose a compression scheme with low memory usage to run on sensor nodes. Their proposed compression scheme, S-LZW improves over an existing compression scheme, LZW compression [58]. This has been achieved by setting desirable attributes for LZW compression on a sensor node with regard to the dictionary size, the data size to compress at one time and the protocol to use when the data dictionary fills up. To further improve the running of LZW algorithm on a sensor node, the authors also proposed the use of an in-memory data cache to filter repetitive sensor data. More recently, in [39], the authors proposed a compression algorithm for WSN that outperforms the S-LZW in terms of compression ratio and computational complexity by exploiting high correlations existing between consecutive data samples collected on the sensor node.

3.2.2   Approximation

In general, the aforementioned data compression techniques focus on reducing the amount of data packets to be transmitted in-network when the accuracy of data collection is important. Alternatively, in applications where approximations in the collected data can be tolerated, approximations on sensory data can be performed instead to further reduce data transmissions. The application of approximation to reduce data transmissions in-network has been explored in [64,54,37,5,35,36,6].

These studies have explored the computation of aggregates in sensor network data. In [64], the authors propose protocols to continuously compute network digests. These digests are specifically digests defined by decomposable functions such as *min, max, average* and *count*. The novelty in their computation of network digests lies in the distributed computation of the digest function at each node in the network in order to reduce communication overhead and promote load-balanced computation. The partial results obtained on the nodes are then piggybacked on neighbour-to-neighbour communication and eventually propagated up to the root node (base station) in the communication tree. On the contrary, in [5], the amount of communications is reduced by having every node store an estimated value of the global aggregate and updates this estimate periodically depending on changes in locally sensed values. The locally stored global aggregate of a node is only exchanged with another neighbourhood node if the aggregate value changes significantly after a local update. A distributed approximation scheme has also been used in [6] in which they proposed a probabilistic grouping algorithm to run on local sensor nodes so that the computed local aggregates can progressively converge to the aggregated value in real-time. The proposed scheme has the additional benefit that it is robust to node link failures. Apart from common computing aggregates such as *min, max, average* and *count* in the discussed approaches, *median* (most frequent data values) is another aggregate function that can be used for gathering sensory data. In this regard, [54] have proposed a distributed data summarization technique, known as *q-digest* that can operate on limited memory. This facilitates computation of aggregates such as medians or histograms.

Similarly, in [37,36], the focus is on reducing the total number of messages required to compute an aggregate in a distributed manner. In [37], the authors propose a scheme termed *pipelined aggregate* in which the aggregates are propagated into the network in time divisions. Applying this scheme, at each time interval, the aggregate is broadcast to sensors one radio hop away. A sensor that hears the request transmit a partial aggregate to its parent by applying the aggregate function to its own readings and readings of its immediate child neighbour. As stated by the authors, the drawback to this scheme is in the number of messages that need to be exchanged to derive the first aggregates from all sensors in the network. In a related work, [36] discuss broader data acquisition issues pertaining to the energy-efficiency of the query dissemination process and energy-optimised query optimisation. For instance, in [36], semantic routing trees are proposed for collecting aggregates from the sensor network.

3.2.3 Prediction

The third class of techniques that can be applied to reduce network data transmissions is prediction. Prediction techniques at the node level derive spatial and temporal relationships or probabilistic models from sensory data to estimate local data readings or readings of neighbouring nodes. When sensory data of particular sensor nodes can be predicted, these sensor nodes are then suppressed from transmitting the data to save communication costs. Similar to compression and approximation, prediction-based techniques are also required to run in a light-weight manner on sensor nodes. In the literature, prediction techniques have been proposed as algorithms for enhancing data acquisition in [34,15,53] and as generic light-weight learning techniques to reduce communication costs in transmissions.

For energy-efficient data acquisition, [34] proposed their Data Stream Management System (DSMS) architecture to optimise the data collection process in querying. In their architecture, sensor proxies are used as mediators between the query processing environment and the physical sensors, where proxies can adjust sensor sampling rates or request for some further operations before sending data by intelligently sampling sensors (for instance, sampling less frequently if the user query demands so) rather than just sampling data randomly. The energy efficiency in data collection has also been addressed in [15], whereby the idea is to select in a dynamic fashion the subset of nodes to perform sensing and transmitting data to the base station and data values for the rest of the nodes predicted using probabilistic models. Similarly, in [53], data to be communicated is reduced by controlling the number of sensor nodes communicating their sensory data. This is achieved by setting threshold values so that a sensor should only send its reading when its reading is outside the set threshold.

In [41], the prediction involves the building of local predictive models at sensor nodes and having sensors transmit the target class predictions to the base station. The models built at sensors can then be used to predict target classes such as to facilitate anomalies detection, which reduces the transmissions of sensor data necessary otherwise to the base station. Such a distributed approach has also been adopted by [13] who propose sensors that re-adjust their actions based on the analysis of information shared among neighbouring sensors. These sensors perform actions that could be more resource and time-efficient. In particular, they acquire spatio-temporal relationships by learning from a neighbourhood of sensor data and history data. Markov models are used to calculate probabilities of the required data fitting into different time intervals. Using the calculated probabilities, sensor readings with the highest confidence for missing sensor data are chosen. Once the correlations between sensor data are learnt and reused over time, the need to send prediction models from the base stations are omitted, thereby saving energy through reduced sensing and communication costs. Experimental results show the efficiency of the classifier based on simulations.

In another study, [10] propose a generalised Non-Parametric Expectation-Maximization (EM) algorithm for sensor networks. Conventionally, a parametric EM algorithm is a clustering algorithm that, from chosen initial values for specified parameters and a probability density function, computes cluster probabilities for each instance. Ultimately, in the maximization step, the algorithm will derive a convergence or local maximum after several iterations. However, for a sensor network, the algorithm is not applicable because sensors frequently report false values, requiring them to calculate the estimates as required by the algorithm, which is infeasible for resource-limited sensor nodes. The solution provided is one that uses the estimation step on nodes such that nodes will estimate their current value from the knowledge of other values from neighbouring nodes. The algorithm requires that sensor nodes report their discrete values. A more recent study is work done by [40] where the authors investigated correlations that can be formed when sensors in loading trucks experience similar vibrations when the trucks send out the same load. The correlation information of the sensor nodes then allowed them to group trucks carrying out the same load. The unique contribution in their work lies in the incremental calculation of the correlation matrix. The idea of exploiting correlations in sensor node readings to reduce transmissions has also been explored in [16].

Generally, the distributed nature of the algorithms proposed in [41,13,40] allow the computation load to be more balanced across the sensor network to achieve computational efficiency. Alternatively, a more centralised node processing model can be applied such as work in [50] whereby the authors propose delegating the base station to calculate the classification model and uploads the predicted model to sensor nodes. Sensor nodes then use the model to selectively report the 'interesting' data points to the base station as a means to reduce energy consumption. Similarly, in Prediction-based Monitoring in Sensor Networks: PREMON [23], the processing is shared by intermediate nodes in the network. In this work, the authors describe a video compression-based prediction technique, the block-matching function of MPEG, to predict spatio-temporal correlation models at

Approach Type	Features	Limitations
Compression-based	+ Preserves the accuracy of readings collected.	- A need for decoding at the base station. - Data compression techniques are computationally heavy.
Approximation-based	+ Low computational overhead relative to data compression.	- Some techniques only applicable to certain WSN applications such as applying aggregates for data querying.
Prediction-based	+ Generic light-weight algorithms have been proposed.	- Some prediction techniques are application-specific. - Prediction may sacrifice data collection accuracy when some nodes are not required to send.

Fig. 5. Data granularity control techniques summary

intermediate nodes. The prediction correlation models are then passed on to sensor nodes within the aggregation group and a sensor node will then send its reading when the readings have not been predicted. To save transmission costs, PRE-MON enables a sensor to only send its actual readings when the readings cannot be inferred from the predicted model. However, the approach has a heavy overhead as predictions often have to be continuously sent to the sensors.

Figure 5 illustrates the qualitative differences between the aforementioned data granularity control techniques.

4   Conclusions

Existing data processing approaches have explored how sensor communication can be improved through utilising processed data obtained from sensors. This has been achieved by utilising a suitable energy trade-off between computation and communication operations:

Network-based approaches. At the network data level, the data model is such that sensor data arrives at the application running at a central processing location, enabling the application to have a global view of data distribution within a sensor network. By utilisation of a resource-rich base station, network-based approaches can utilise information obtained from expert knowledge or prediction at network data level to obtain cues that would improve sensing and communication operations at a high level.

Node-based approaches. At the node data level, as the computation process is performed locally on sensor nodes, efforts towards energy conservation focus on developing data processing algorithms suitable for resource-constrained sensor nodes. This involves optimising the data processing algorithm to operate on minimal storage using limited processing power while being adaptive to the sensor's remaining energy. Through local data processing data on sensor nodes, the overall amount of data that needs to be transmitted in-network is then consequently reduced, i.e. applying either compression, approximation or prediction.

It can be observed that approaches at the network level have addressed energy conservation by controlling certain aspects of network operation. The work done thus far has focused on improving particular node operations, for instance in sampling, the aim is to reduce the number of nodes in a group that send data to the base station and consequently reduce energy consumption. Similarly, for node level approaches, energy conservation has focussed on reducing data (lossy or lossless) sent from nodes to the base station through building light-weight processing algorithms for sensor networks using compression, approximation or prediction.

Underlying these approaches, we can observe the notion that if some information about the sensor network can be obtained, then this information can be used to drive sensor network operations efficiently. In the aforementioned approaches, this notion has not been explicated or formalised. Therefore, one direction of future work could be to explore using computed information (locally at sensor node or globally at the base station) to autonomously decide how a sensor should operate efficiently at any point during its sensing task. For example, if the computed information suggests that sensing is no longer required at a sensed region, then the energy-efficient operation to be carried out by sensors in that sensed region may be to sleep or to sense less frequently. In addition, this raises related research questions, namely, how then would streaming sensory information be obtained efficiently and how it would be used to control sensors in a scalable manner.

References

1. Akyildiz, I., Su, W., Sankarasubramaniam, Y., Cayirci, E.: Wireless sensor networks: A survey. Computer Networks: The International Journal of Computer and Telecommunications Networking 38(4), 393–422 (2002)
2. Anastasi, G., Conti, M., Falchi, A., Gregori, E., Passarella, A.: Performance measurements of mote sensor networks. In: ACM/IEEE International Symposium on Modeling, Analysis and Simulation of Wireless and Mobile Systems, Venice, Italy, pp. 174–181 (2004)
3. Arici, T., Gedik, B., Altunbasak, Y., Liu, L.: Pinco: A pipelined in-network compression scheme for data collection in wireless sensor networks. In: 12th International Conference on Computer Communications and Networks, Texas, USA, pp. 539–544 (2003)
4. Barr, K., Asanovic, K.: Energy-aware lossless data compression. ACM Transactions on Computer Systems 24(3), 250–291 (2006)
5. Boulis, A., Ganeriwal, S., Srivastava, M.: Aggregation in sensor networks: An energy-accuracy tradeoff. In: 1st IEEE International Workshop on Sensor Network Protocols and Applications, California, USA, pp. 128–138 (2003)
6. Chen, J., Pandurangan, G., Xu, D.: Robust computation of aggregates in wireless sensor networks: Distributed randomized algorithms and analysis. In: 4th International Symposium on Information Processing in Sensor Networks, California, USA, pp. 348–355 (2005)
7. Chou, J., Petrovic, D., Ramchandran, K.: A distributed and adaptive signal processing approach to reducing energy consumption in sensor networks. In: IEEE International Conference of the IEEE Computer and Communications Societies, San Francisco, USA, pp. 1054–1062 (2003)
8. Chu, D., Deshpande, A., Hellerstein, J., Hong, W.: Approximate data collection in sensor networks using probabilistic models. In: 22nd International Conference on Data Engineering, Atlanta, USA (2006)
9. Dalton, A., Ellis, C.: Sensing user intention and context for energy management. In: 9th Workshop on Hot Topics in Operating Systems, USENIX, Hawaii, USA, pp. 151–156 (2003)
10. Davidson, I., Ravi, S.: Distributed pre-processing of data on networks of berkeley motes using non-parametric em. In: 1st Internation Workshop on Data Mining in Sensor Networks, Los Angeles, USA, pp. 17–27 (2005)

11. Deshpande, A., Guestrin, C., Madden, S., Hellerstein, J., Hong, W.: Model-driven data acquisition in sensor networks. In: 30th Very Large Data Base Conference, Toronto, Canada, pp. 588–599 (2004)
12. Deshpande, A., Madden, S.: Mauvedb: Supporting model-based user views in database systems. In: Special Interest Group on Management of Data, Illonois, USA, pp. 73–84 (2006)
13. Elnahrawy, E., Nath, B.: Context-aware sensors. In: Karl, H., Wolisz, A., Willig, A. (eds.) EWSN 2004. LNCS, vol. 2920, pp. 77–93. Springer, Heidelberg (2004)
14. Gaber, M., Krishnaswamy, S., Zaslavsky, A.: Mining data streams: A review. ACM SIGMOD Record 34(2), 18–26 (2005)
15. Gedik, B., Liu, L., Yu, P.: Asap: An adaptive sampling approach to data collection in sensor networks. IEEE Transactions on Parallel and Distributed Systems 18(12), 1766–1782 (2007)
16. Goel, S., Passarella, A., Imielinski, T.: Using buddies to live longer in a boring world. In: 4th Annual IEEE International Conference on Pervasive Computing and Communications, Washington, USA, p. 342 (2006)
17. He, T., Krishnamurthy, S., Stankovic, J., Abdelzaher, T., Luo, L., Stoleru, R., Yan, T., Gu, L., Hui, J., Krogh, B.: Energy-efficient surveillance system using wireless sensor networks. In: 2nd International Conference on Mobile Systems, Applications and Services, Boston, USA, pp. 270–283 (2004)
18. Hefeeda, M., Bagheri, M.: Wireless sensor networks for early detection of forest fires. In: IEEE International Conference on Mobile Adhoc and Sensor Systems, Pisa, Italy, pp. 1–6 (2007)
19. Heinzelman, W., Chandrakasan, A., Balakrishnan, H.: Energy-efficient communication protocol for wireless microsensor networks. In: 33rd Annual Hawaii International Conference on System Sciences, Maui, USA, pp. 2–12 (2000)
20. Heinzelman, W., Chandrakasan, A., Balakrishnan, H.: An application-specific protocol architecture for wireless microsensor networks. IEEE Transactions on Wireless Communications 1(4), 660–670 (2002)
21. Hoang, A., Motani, M.: Exploiting wireless broadcast in spatially correlated sensor networks. In: International Conference on Communications, Seoul, Korea, pp. 2807–2811 (2005)
22. Hu, W., Tran, V., Bulusu, N., Chou, C., Jha, S., Taylor, A.: The design and evaluation of a hybrid sensor network for cane-toad monitoring. In: 4th International Symposium on Information Processing in Sensor Networks, California, USA, pp. 28–41 (2005)
23. Imielinski, T., Goel, S.: Prediction-based monitoring in sensor networks: Taking lessons from mpeg. ACM Computer Communication Review 31(5), 82–98 (2001)
24. Intanagonwiwat, C., Estrin, D., Govindan, R., Heidemann, J.: Impact of network density on data aggregation in wireless sensor networks. In: 22nd International Conference on Distributed Computing Systems, Vienna, Austria, p. 457 (2002)
25. Intanagonwiwat, C., Govindan, R., Estrin, D.: Directed diffusion: A scalable and robust communication paradigm for sensor networks. In: ACM/IEEE International Conference on Mobile Computing and Networking, Boston, USA, pp. 56–67 (2000)
26. Jain, A., Chang, E., Wang, Y.: Adaptive stream resource management using kalman filters. In: Special Interest Group on Management of Data, Paris, France, pp. 11–22 (2004)
27. Kamimura, J., Wakamiya, N., Murata, M.: A distributed clustering method for energy-efficient data gathering in sensor networks. International Journal on Wireless and Mobile Computing 1(2), 113–120 (2006)

28. Kanagal, B., Deshpande, A.: Online filtering, smoothing and probabilistic modeling of streaming data. In: 24th International Conference on Data Engineering, Cancun, Mexico, pp. 1160–1169 (2008)
29. Keshavarz, A., Tabar, A.M., Aghajan, H.: Distributed vision-based reasoning for smart home care. In: ACM SenSys Workshop on Distributed Smart Cameras, Boulder, USA, pp. 105–109 (2006)
30. Kimura, N., Latifi, S.: A survey on data compression in wireless sensor network. In: International Conference on Information Technology: Coding and Computing, Washington, USA, pp. 8–13 (2005)
31. Krishnamachari, B., Estrin, D., Wicker, S.: The impact of data aggregation in wireless sensor networks. In: 22nd International Conference on Distributed Computing Systems, Washington, USA, pp. 575–578 (2002)
32. Kumar, R., Tsiatsis, V., Srivastava, M.: Computation hierarchy for in-network processing. In: 2nd ACM International Conference on Wireless Sensor Networks and Applications, California, USA, pp. 68–77 (2003)
33. Lindsey, S., Raghavendra, C.: Pegasis: Power-efficient gathering in sensor information systems. In: Aerospace Conference Proceedings, Montana, USA, pp. 1125–1130 (2002)
34. Madden, S., Franklin, M.: Fjording the stream: An architecture for queries over streaming sensor data. In: 18th International Conference on Data Engineering, San Jose, USA, pp. 555–566 (2002)
35. Madden, S., Franklin, M., Hellerstein, J., Hong, W.: Tag: a tiny aggregation service for ad hoc sensor networks. In: 5th Annual Symposium on Operating Systems Design and Implementation, Boston, USA, pp. 131–146 (2002)
36. Madden, S., Franklin, M., Hellerstein, J., Hong, W.: The design of an acquisitional query processor for sensor networks. In: ACM Special Interest Group on Management of Data, Wisconsin, USA, pp. 491–502 (2003)
37. Madden, S., Szewczyk, R., Franklin, M., Culler, D.: Supporting aggregate queries over ad-hoc wireless sensor networks. In: 4th IEEE Workshop on Mobile Computing Systems and Applications, New York, USA, pp. 49–58 (2002)
38. Manjhi, A., Nath, S., Gibbons, P.: Tributaries and deltas: Efficient and robust aggregation in sensor network stream. In: Special Interest Group on Management of Data, Baltimore, USA, pp. 287–298 (2005)
39. Marcelloni, F., Vecchio, M.: A simple algorithm for data compression in wireless sensor networks. IEEE Communications letters 12(6), 411–413 (2008)
40. Marin-Perianu, R., Marin-Perianu, M., Havinga, P., Scholten, H.: Movement-based group awareness with wireless sensor networks. In: Pervasive, Toronto, Canada, pp. 298–315 (2007)
41. McConnell, S., Skillicorn, D.: A distributed approach for prediction in sensor networks. In: 1st International workshop on Data Mining in Sensor Networks, Newport Beach, USA, pp. 28–37 (2005)
42. Mini, R., Nath, B., Loureiro, A.: A probabilistic approach to predict the energy consumption in wireless sensor networks. In: IV Workshop de Comunicacao sem Fio e Computacao Movel, Sao Paulo, Brazil, pp. 23–25 (2002)
43. Nath, S., Gibbons, P., Seshan, S., Anderson, Z.: Synopsis diffusion for robust aggregation in sensor networks. In: ACM Conference on Embedded Networked Sensor Systems, Baltimore, USA, pp. 250–262 (2004)
44. Passos, R., Nacif, J., Mini, R., Loureiro, A., Fernandes, A., Coelho, C.: System-level dynamic power management techniques for communication intensive devices. In: International Conference on Very Large Scale integration, pp. 373–378. Nice, French Riviera (2006)

45. Pattem, S., Krishnamachari, B., Govindan, R.: The impact of spatial correlation on routing with compression in wireless sensor networks. ACM Transactions on Sensor Networks 4(4), 24–33 (2008)
46. Perillo, M., Ignjatovic, Z., Heinzelman, W.: An energy conservation method for wireless sensor networks employing a blue noise spatial sampling. In: International Symposium on Information Processing in Sensor Networks, California, USA, pp. 116–123 (2004)
47. Petrovic, D., Shah, R., Ramchandran, K., Rabaey, J.: Data funneling: routing with aggregation and compression for wireless sensor networks. In: 1st IEEE International Workshop on Sensor Network Protocols and Applications, California, USA, pp. 156–162 (2003)
48. Pottie, G., Kaiser, W.: Wireless integrated network sensors. Communications of the ACM 43(5), 51–58 (2000)
49. Puri, A., Coleri, S., Varaiya, P.: Power efficient system for sensor networks. In: 8th IEEE International Symposium on Computers and Communication, Kemer, Antalya, Turkey, vol. 2, pp. 837–842 (2003)
50. Radivojac, P., Korad, U., Sivalingam, K., Obradovic, Z.: Learning from class-imbalanced data in wireless sensor networks. In: 58th Vehicular Technology Conference, Florida, USA, vol. 5, pp. 3030–3034 (2003)
51. Sadler, C., Martonosi, M.: Data compression algorithms for energy-constrained devices in delay tolerant networks. In: ACM Conference on Embedded Networked Sensor Systems, Colorado, USA, pp. 265–278 (2006)
52. Seward, J.: bzip2 compression algorithm (2008), http://www.bzip.org/index.html
53. Sharaf, M., Beaver, J., Labrinidis, A., Chrysanthis, P.: Tina: A scheme for temporal coherency-aware in-network aggregation. In: ACM Workshop on Data Engineering for Wireless and Mobile Access, California, USA, pp. 69–76 (2003)
54. Shrivastava, N., Buragohain, C., Agrawal, D., Suri, S.: Medians and beyond: New aggregation techniques for sensor networks. In: ACM Conference on Embedded Networked Sensor Systems, Baltimore, USA, pp. 239–249 (2004)
55. Tian, D., Georganas, N.: A node scheduling scheme for large wireless sensor networks. Wireless Communications and Mobile Computing Journal 3(2), 271–290 (2003)
56. Tulone, D., Madden, S.: Paq: Time series forecasting for approximate query answering in sensor networks. In: 3rd European Conference on Wireless Sensor Networks, Zurich, Switzerland, pp. 21–37 (2006)
57. Vuran, M., Akyildiz, I.: Spatial correlation-based collaborative medium access control in wireless sensor networks. IEEE/ACM Transactions on Networking 14(2), 316–329 (2006)
58. Welch, T.: A technique for high-performance data compression. IEEE Computer 17(6), 8–19 (1984)
59. Willett, R., Martin, A., Nowak, R.: Backcasting: Adaptive sampling for sensor networks. In: International Symposium on Information Processing in Sensor Networks, California, USA, pp. 124–133 (2004)
60. Ye, F., Zhong, G., Cheng, J., Lu, S., Zhang, L.: Peas: A robust energy conserving protocol for long-lived sensor networks. In: 23rd International Conference on Distributed Computing Systems, Providence, Rhode Island, pp. 28–37 (2003)
61. Ye, M., Li, C., Chen, G., Wu, J.: Eecs: An energy efficient clustering scheme in wireless sensor networks. In: 24th IEEE International Performance Computing and Communications Conference, Arizona, USA, pp. 535–540 (2005)

62. Yoon, S., Shahabi, C.: The clustered aggregation (cag) technique leveraging spatial and temporal correlations in wireless sensor networks. ACM Transactions on Sensor Networks 3(1), 1–39 (2007)
63. Younis, O., Fahmy, S.: Heed: A hybrid, energy-efficient, distributed clustering approach for ad-hoc sensor networks. IEEE Transactions on Mobile Computing 3(4), 366–379 (2004)
64. Zhao, J., Govindan, R., Estrin, D.: Computing aggregates for monitoring wireless sensor networks. In: 1st IEEE International Workshop on Sensor Network Protocols and Applications, California, USA, pp. 139–148 (2003)

Improving Source Selection
in Large Scale Mediation Systems
through Combinatorial Optimization Techniques

Alexandra Pomares[1], Claudia Roncancio[2],
Van-Dat Cung[2], and María-del-Pilar Villamil[3]

[1] Pontificia Universidad Javeriana, Bogotá, Colombia
[2] Grenoble INP, Grenoble, France
[3] Universidad de los Andes, Bogotá, Colombia

Abstract. This paper concerns querying in large scale virtual organizations. Such organizations are characterized by a challenging data context involving a large number of distributed data sources with strong heterogeneity and uncontrolled data overlapping. In that context, data source selection during query evaluation is particularly important and complex. To cope with this task, we propose OptiSource, an original strategy for source selection using combinatorial optimization techniques combined to organizational knowledge of the virtual organization. Experiment numerical results show that OptiSource is a robust strategy that improves the precision and the recall of the source selection process. This paper presents the data and knowledge models, the definition of OptiSource, the related mathematical model, the prototype and an extensive experimental study.

Keywords: Large Scale Data Mediation, Source Selection, Combinatorial Optimization.

1 Introduction

Source Selection (also known as server or database selection) is the process of identifying the set of data sources that must participate in the evaluation of a query. Source selection is particularly critical in large scale systems where the number of possible query plans grows rapidly as the complexity of the query and the set of data sources increases. Research in this area has been promoted mainly by the requirements of the world wide web, distributed file sharing systems and distributed databases.

This paper explores the problem of source selection in large scale data contexts that are typical of the new type of organizational model called Virtual Organization (VO). They combine characteristics of web data contexts, data heterogeneity and high distribution, and characteristics of enterprise data contexts, like availability commitment and structured data sources.

A. Hameurlain, J. Küng, and R. Wagner (Eds.): TLDKS III , LNCS 6790, pp. 138–166, 2011.

1.1   Querying in Large Scale Virtual Organizations

A Virtual Organization (VO) is a set of autonomous collaborating organizations, called VO units, working towards a common goal. It enables disparate groups to share competencies and resources such as data and computing resources [1]. VOs work in a domain (e.g. health) and their participants share information about objects of interest in the domain (e.g. patients, clients). Instances of those types of objects have to be composed from fragments distributed in the data sources of the VO participants. Data may be partially replicated on several sources. Considering a type of object, an important type of query is the selection of the instances satisfying a given condition (a selection predicate). For example, considering the type *Patients*, a query seeking those such as *gender = female and diagnosis = cancer* may potentially involve querying all data sources having a fragment of a Patient. However, if a complete answer is not required (e.g. the user is doing a research that does not need all the patients, but a good sample), the query optimizer may use heuristics to prune some data sources in order to speed up the response time of the query.

1.2   Challenges in Source Selection in VO Data Contexts

Source selection and data integration are particularly challenging in VO data contexts. As VOs have evolved to national and world-wide magnitudes [2,3], it is unlikely that VO participants provide precise metadata describing the content of their data sources and if they do maintain such metadata, it would be costly. Furthermore, due to the **heterogeneity** of data sources, data may be provided with **different levels of details** making source selection harder. For example, to look for patients having a *cancer*, the system has to be able to recognize that sources having data of patients with *breast cancer* or *invasive lobular carcinoma* are relevant. A deep analysis between metadata and queries is therefore required to come out with an effective query plan.

Another challenge is introduced by the **uncontrolled data overlapping** between data sources. An instance of a subject may exist in several data sources which may or may not have overlapping sets of properties. Information may be **replicated** without having necessarily "well formed copies". This prevents the use of strategies that consider data sources are complete or independent [4].

The aforementioned characteristics of the data context may be a consequence of business processes in the VO. **Heterogeneity, source overlapping, large distribution and a high number of data sources** make query processing a complex task, specially in the planning phase. Several existing proposals on source selection take into account distribution, heterogeneity and integration of structured data sources [5,6,7,8] . Nevertheless, as shown in the following, they are not well suited for VOs, some of them can be very inefficient or simply cannot be used in VO data contexts. The main reasons are that such proposals do not address uncontrolled replication and non-disjoint fragments, or because their requirements of data source metadata are not fulfilled in VOs.

1.3   Contribution and Paper Organization

The work presented in this paper is a contribution to improve source selection in large scale virtual organizations having a complex data context. It presents OptiSource, a source selection strategy combining two original aspects:

- The first one is the introduction of roles for data sources based on knowledge associated to the VO. A seamless integration of such roles with the taxonomy of concepts used in the VO permits a good estimation of the relevance of using a data source to evaluate a query.
- The second one is the use of a combinatorial optimization model to control the number of sources to contact. The objective function maximizes the benefit of using a data source and minimizes the number of sources to contact while satisfying processing resource constraints. The efficiency of source selection is improved so as its recall and precision criteria.

To the best of our knowledge there are not yet other proposals using combinatorial optimization techniques in conjunction with ontologies during query evaluation. Our research shows that this approach is promising.

This paper presents the full proposal of OptiSource, its definition, a prototype and a performance study. It is an extension of the work presented in [9]. It includes a deep analysis of related works, it provides a detailed presentation of each component of OptiSource, and it analyzes the complexity of the mathematical model used to optimize the assignment of data sources to query conditions. In addition, it presents in detail the experimental environment and reports additional tests and analyses about the impact of prediction accuracy during the process of source selection.

In the following, Section 2 presents the terminology and notation used in the paper. Section 3 gives an analysis of related works in source selection and the utility of existing proposals in large scale VOs. Section 4 presents OptiSource, the source selection strategy we propose. Section 5 presents the combinatorial optimization model used in OptiSource. Section 6 presents the OptiSource prototype. A large variety of tests have been executed to evaluate our proposal. The results are discussed in Sections 6 and 7. Finally, Section 8 presents our conclusions and research perspectives.

2   Preliminaries

The source selection strategy proposed in this work is part of a mediation system for large scale VOs. At the mediation level data are represented as Virtual Data Objects (VDO) which link together several concepts.

Definition 1 *Virtual Data Object (VDO). A VDO is a logical set of related concepts relevant to a group of users. Figure 1 illustrates a VDO joining four concepts. Each concept has data type properties whose values are data literals. Concepts are related through object type properties whose values are instances. VDOs are virtual because their instances are partitioned and distributed on several data sources.*

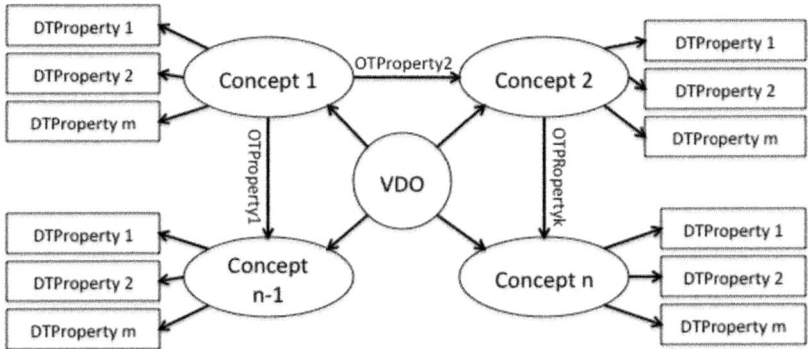

Fig. 1. Virtual Data Object

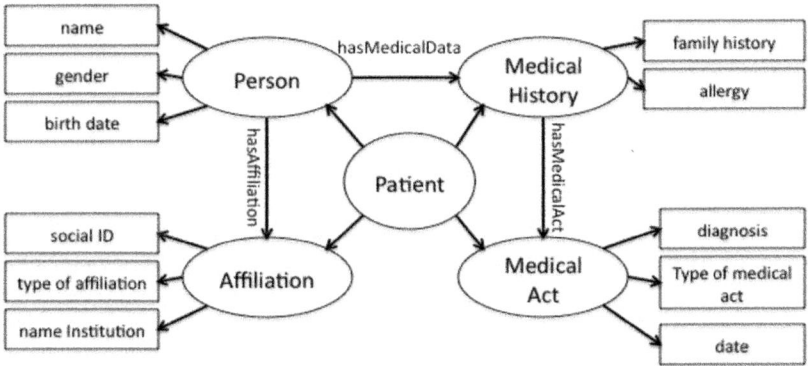

Fig. 2. VDO Patient

Figure 2 shows a VDO *Patient*. Four concepts (Person, Medical Act, Medical History, Affiliation) are related to compose the VDO *Patient*.

We consider here queries with selection conditions on the properties of the concepts composing the VDO. For example, the query hereafter uses the *VDO Patient* and selects the id of patients with a "Buphtalmos" diagnosis.

Query 1. $Q(VDO_{Patient}, properties(id), conditions(diagnosis = Buphtalmos))$

In the following we will work with queries including several conditions as follows:

$$Q(VDO_{name}, properties(id), conditions(condition_1, condition_2, ..., condition_n))$$

Data abstraction levels

We consider an enriched architecture of mediation with three levels of data abstraction (see Figure 3 for an example). The external level presents the set of VDOs supported by the system. Queries will refer to them. The conceptual level contains the reference ontology that defines the domain of the VO. The concepts defined in this ontology are used to create VDOs. Finally, the internal

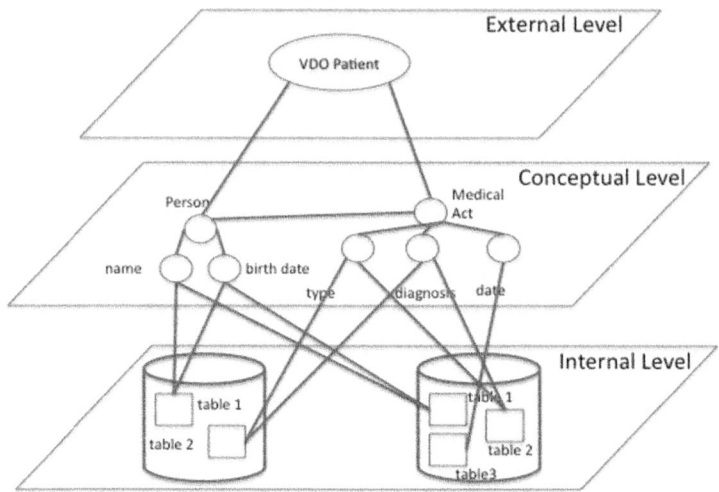

Fig. 3. Data Abstraction Levels

level contains all the data sources available in the VO. Such data is used to build the instances of the VDO concepts.

The rest of the paper will be focused on the strategy used to identify the data sources from the internal level that must be used to evaluate the query defined at the external level.

3 Data Source Selection: Related Works

Even though multi-sources querying has been intensively studied for more than 15 years, none of the available strategies for query planning can be directly applied for large scale VO data contexts.

This section presents an analysis of the main source selection strategies available in the literature in order to identify interesting ideas that could be applied on large scale VO contexts. We analyze several proposals related to source selection including structured, semi-structured and non structured sources as seen in Figure 4. However, in this paper we consider the most representative proposals such as capability, multi-utility and quality oriented systems.

3.1 Capability Oriented Systems

In the first generation of integration systems for structured sources like Information Manifold [10], TSIMMIS [11], DISCO [12] and Minicon [8], source selection is performed according to the capabilities of sources. These capabilities are associated to the capacity to evaluate types of queries, for example the number of conditions and the attributes that can be restricted in a query. In this way, they require knowledge about the intentional level and the computing capacity [13] (e.g. which attributes can be used as filters) of each data source.

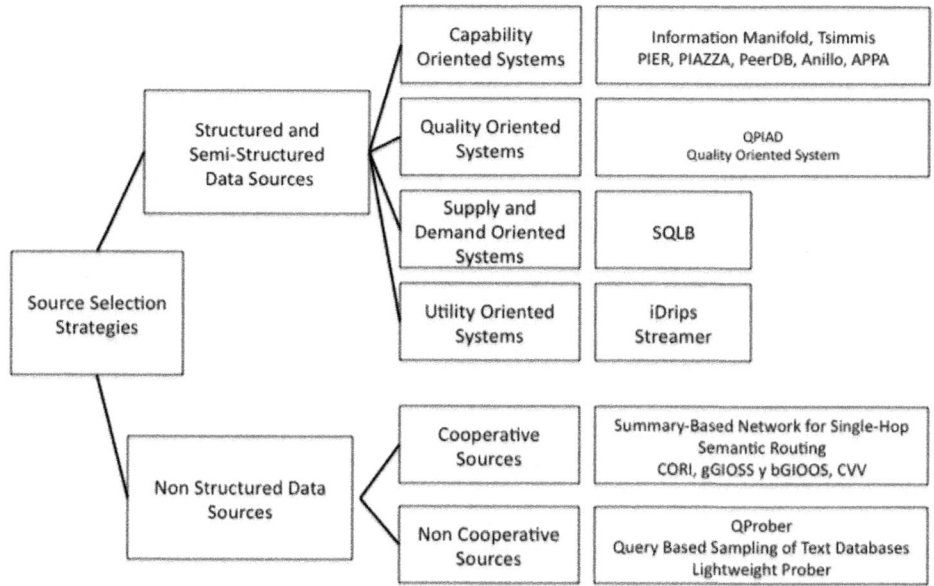

Fig. 4. Source Selection Strategies

Although such proposals are appropriate in several contexts, they are not able to reduce significantly the set of data sources that must participate to evaluate a query in a VO because the processing capabilities of sources are typically alike and, as a consequence, are not useful to differentiate data sources.

3.2   Multi-utility Oriented Systems

Another group of proposals like iDrips and Streamer [5] conceive the source selection as the identification of most useful sources for executing a query. These sources are selected comparing utility functions of different plans for executing the query.

iDrips and Streamer assume source similarity property. Source Similarity means that similar sources can be replaced one by another without affecting the utility of a plan. Whereas Plan Independence means that two plans are independent if the utility of a plan does not depend on whether the other has been executed.

iDrips algorithm acts in two phases: the first phase groups similar data sources and creates abstract plans that are capable of solving all the predicates of the query. Each abstract plan has an utility and represents a set of concrete plans.

The second phase obtains query plans in decreasing order of utility. It is decomposed into iterations. Each iteration selects the most dominant abstract plan, removes it from the plans space, and rebuilds the dominance relationships that are necessary. A dominant abstract plan contains at least one concrete plan that maximizes the utility in comparison to all concrete plans in other abstract plans.

The Streamer algorithm is a modification of iDrips. Streamer assumes that plans are independent avoiding to recalculate the dominance relationships in each iteration.

The effectiveness of these algorithms depends on the utility function. Several utility functions are proposed to represent the relationships between data sources. However, with the coverage utility function, used in the paper [5], it is not clear how to obtain variables involved in its calculation. Additionally, even though the coverage of query plans could be obtained, it does not consider the extensional overlapping between sources incurring on inefficiencies during the ordering. This lead to restrict the use of the algorithm on VO contexts where utility functions must be able to express relationships between sources like overlapping and replication.

3.3   Quality Oriented Systems

More recent works like the strategy proposed in Bleiholder et al. [14] assume the source selection as the identification of the better paths between overlapped data sources to resolve a query. These paths maximize the benefit in terms of the number of objects obtained in the answer, while respecting a cost constraint or budget limitations.

They model the problem of selecting the best set of paths as a dual problem of budgeted maximum coverage (BMC) and maximal set cover with a threshold (MSCT). These problems were solved through three types of algorithms: integer programming/linear programming algorithm, greedy algorithms and a random search algorithm.

According to BMC problem, greedy algorithms are useful only with higher budgets; whereas linear programming algorithms and the random algorithm have good behavior with different levels of budgets. The MSCT problem is solved only with greedy algorithms, which do not always find the optimal solution. This proposal is very efficient in data contexts where it is possible to know the instances that are contained in each data source. However, assumptions of this kind of detailed knowledge about data sources is unavailable in most of the distributed data contexts. Consequently, the application of the proposed model in this kind of contexts is not feasible. The strategies to obtain the relationships between objects of data sources and the benefit of each path should be provided in order to enable the use of these algorithms in any context.

Other proposals like QPIAD [6] and Quality Oriented system [15] use detailed statistics on data source quality to reduce the number of possible query plans and therefore sources. QPIAD is a rewriting strategy that allows to retrieve relevant answers from autonomous data sources, using correlations between attributes used on predicates of a query. These answers are incomplete because they include answers obtained using query rewritings where some conditions will be replaced, in particular using *null* values. For instance, a query with model=Accord and make=Honda conditions can be rewriting as model=Accord and make=null, generating incomplete answers.

QPIAD principle is to learn attribute correlation (e.g. make and model), value distribution (e.g. Honda and Accord) and query selectivity in order to retrieve incomplete answers that have a high probability of being certain answers.

The QPIAD algorithm acts in two phases: the first one sends the original query to all the available data sources to obtain the set of certain answers. These answers are sent to the final user. The second phase learns approximate correlations between attribute values, using the set of certain answers and a sample of data sources that was previously built. These correlations are used to generate the rewritten queries that are sent to data sources to obtain possible answers. Although these queries may retrieve possible answers they do not have the same relevance. Consequently, the algorithm ranks the queries using their estimated precision and recall, and then QPIAD sends them to the data sources. Only the first k queries are sent to the data sources. The value of k is the maximum number of queries supported by each data source.

QPIAD allows to obtain complete answers in data contexts where data sources have quality problems. This strategy includes an innovative algorithm to learn approximate functional dependencies between attribute values that can be useful not only in source selection problems, but also in data mining projects. However, the rewriting phase of QPIAD in large scale distributed data contexts leads to scalability problems. Even though the queries defined by QPIAD improves the quality of the result, it is heavily resource dependent. The algorithm of ranking of QPIAD can include a phase of analysis of data sources in order to predict which data sources are more relevant for the query.

Finally, proposals using a P2P approach (PIER [7], PIAZZA [16], EDUTELLA [17], SomeWhere [18]) manage numerous sources. However, as their rewriting and source selection strategies do not take into account source overlapping, they would lead to high redundancy in the answers when used in large scale VOs.

4   OptiSource: A Decision System for Source Selection

The analysis of related works led us to identify the lack of a strategy that assures scalability of the query evaluation w.r.t. the number of available data sources when there are overlapping and uncontrolled replication between data sources. This gap, critical for the mediation in large scale data contexts, is fulfilled with OptiSource, a strategy of source selection created for large scale VOs. This section presents OptiSource, its principle and its general process in Section 4.1 and its components in Sections 4.2, 4.3, 4.4, 4.5.

4.1   Principle and General Process

In order to enable scalability, OptiSource principle is to identify for each condition the minimal set of sources required to obtain the maximum number of instances that match the condition, while avoiding the contact of unnecessary sources. The minimal set of data sources is identified selecting iteratively the most profitable data sources w.r.t. the query. The profit of data sources w.r.t. a query reflects the estimated benefit of using a data source to evaluate a query and is calculated using the aspects presented in Table 1. The data sources whose estimation of benefit is higher are considered the most profitable. The first iteration selects the data sources with higher benefit w.r.t. each condition of the query. The subsequent iterations remove the sources selected in previous iterations and identify the next most profitable data sources. The iterations continue until the user is satisfied with the number of obtained instances or until there is no more available data sources.

Table 1. Profitability aspects

Aspect	Description
Intentional matching	Identify which sources contain VDO properties required in the query
Extensional matching	Identify which data sources contain instances that match all or part of the predicate of the query, and from these which will better contribute in terms of instances that match the predicate
Data Source compatibility	Due to the incompleteness of data sources, in order to avoid efforts of integration of data sources that do not contain instances in common, it must identify the sets of data sources whose join operation will not produce an empty set of instances
Data Source redundancy	Identify which data sources can be filtered due to its redundancy in terms of instances with other sources

The principle of OptiSource is materialized in an iterative process whose inputs are the n available data sources DSi with their mapping (or view) M_i to the VDO (e.g. *Person, MedicalAct*) and the conditions p_j involved in the query predicate:

$$Input = [(DS1, M_1), (DS2, M_2), ..., (DSn, M_n)], [p_1, p_2, ..., p_m].$$

The output of the process is the collection of most profitable joining sets. A joining set is defined as a group of data sources able to evaluate all the conditions of the query predicate and whose join operation between the data sources it comprises will not produce an empty set of instances.

$$Output = (JS_1, JS_2, ..., JS_p).$$

where $JS_i = (DS_k, p_1), ..., (DS_l, p_m), 1 \leqslant k, l \leqslant n$. A couple (DS_k, p_j) means that the data source DS_k will evaluate the condition p_j. A source can evaluate several conditions p_j and a condition p_j can be evaluated by several sources.

Figure 5 presents the general activities of the process conducted by OptiSource. First, OptiSource queries a dynamically updated knowledge base that

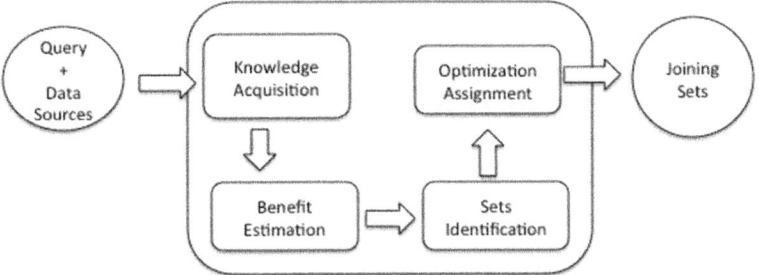

Fig. 5. Activities of OptiSource

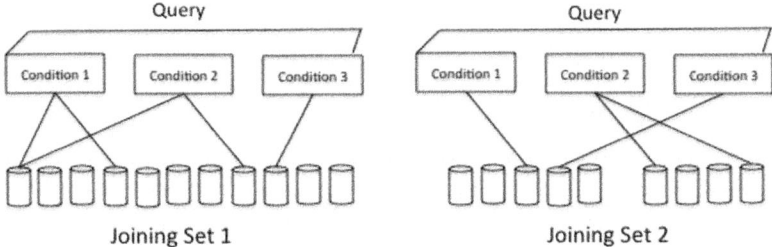

Fig. 6. OptiSource Output

contains the facts describing the characteristics of data sources. With the acquired knowledge it estimates the benefit of using each data source to evaluate the query and identifies the sets of data sources with high tendency to share instances. Next, using the estimated benefit it optimizes the assignment of conditions within each set of data sources. At the end, the output are the joining sets identifying the conditions that each data source will evaluate. Figure 6 presents an example of the output of the process.

4.2   Components of OptiSource

The activities of OptiSource are executed by the components presented in Figure 7. The *Selection Coordinator* is in charge of controlling the execution of the activities to produce the output of the source selection process. The *Knowledge Administrator* is responsible for controlling the interaction with the *VO Knowledge Base*, it queries the base and provides the other components with the knowledge facts it obtains. The *Benefit Estimator* uses the *VO Knowledge Base* facts to predict the benefit of using a data source to evaluate the conditions of a query predicate The *Joining Set Manager* determines the sets of data sources that can work together in order to evaluate the query. Finally, the *Assignment Optimizer* optimizes the assignment of each condition within each joining set. The rest of this section and Section 5 explain the logic of these components.

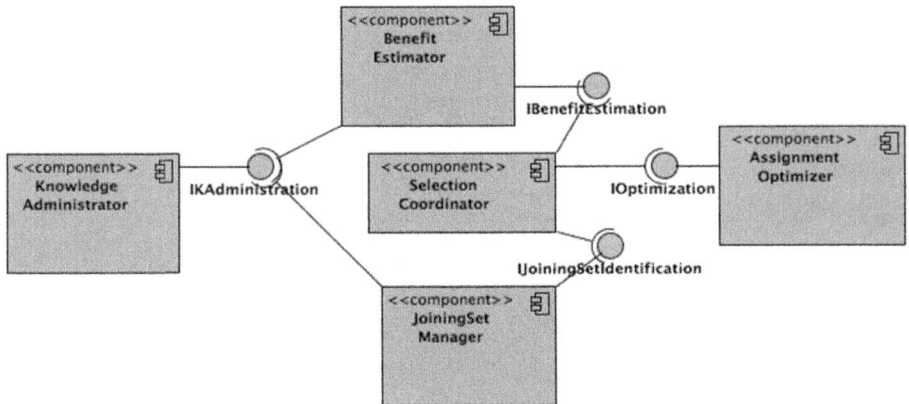

Fig. 7. OptiSource Components

4.3   Knowledge Administrator

The *Knowledge Administrator* controls the interaction with the *VO Knowledge Base*, which is the component of the mediation system that maintains knowledge facts and events related to the data context of the VO. This includes intentional (schemas) and extensional information (contents) of sources. The knowledge is represented as an ontology in OWL [19] with three initial classes: VOUnit, VOResource and VODomainConcept. These classes are specialized and related to better reflect the evolution of the VO data context. **VOUnit** class represents the participants of the VO. An instance of VOUnit can be atomic or composite. In the latter case, it represents a temporal or a permanent group of atomic VO units working together around a specific collaboration process. **VOResource** class represents the physical or logical resources (like data sources) provided by VO units. Finally, **VODomainConcept** includes the subclasses that describe the domain of the VO. For a health VO for instance, initial classes are *Patient*, *Medical Act*, and so on.

4.4   Benefit Estimator Using Data Source Roles

OptiSource estimates the benefit (see Definition 2) of using a source in the evaluation of a query to identify the level of relevance it will have in the query. *DSa* is more relevant than *DSb* if it provides more instances that match the query predicate, even if both of them have the same schema. In order to estimate the benefit we use the available knowledge about the extension (contents) of sources. To obtain this knowledge we work under the assumption that the organizational dimension of VOs allows to relate the role that VO units play in the VO with the contents of the sources they provide. The more complete and reliable the extensional knowledge is, the more accurate the measurement of the benefit is.

Definition 2 *Data Source Benefit.* *Given a query Q with a query predicate P with conditions $[p_1, p_2, ..., p_m]$ the benefit of using a data source DSi to evaluate Q is a variable that measures the degree of contribution of DSi in terms of instances that match one or more $p_j \in P, 1 \leqslant j \leqslant m$.*

Data source roles. In a health VO, for instance, two data sources DSa and DSb provide information about procedures performed on patients. Let us assume that DSa is provided by a VO unit specialized on cancer whereas DSb is provided by a hospital specialized on pediatric procedures. In this case, the DSa may be considered specialist of patients with cancer (vo:Patient, hasDiagnosis vo:Cancer) whereas the DSb is children specialist (vo:Patient, hasAge $\leq$ 15). A source can therefore play different **roles** as a contributor of instances of a VDO (e.g, vo:Patient) verifying some predicate conditions (e.g, hasAge $\leq$ 15). Roles reflect the ability of sources to solve conditions. Given the analysis of the roles played by VO units in the business processes of the VO, we propose the following three roles for their data sources: *authority*, *specialist* and *container*.

The definition of each source role is described in Definition 3, 4 and 5. In these definitions all the instances of $VDOj$ stored in data source DSi are noted $ext(DSi, VDOj)$[1]. U designates the data sources participating in the VO. All the instances of $VDOj$ available in the VO are denoted $ext(U, VDOj)$. The subset of $ext(DSk, VDOj)$ corresponding to the instance that verifies a condition p is denoted $ext(DSk, VDOj)^p$ and $card()$ is the cardinality function.

Definition 3 Authority Role. *A data source DSi plays an authority role w.r.t. a condition p in a query on $VDOj$ iff it stores all the instances of $VDOj$ available in U that match p. $IsAuthority(DSi, VDOj, p) \implies ext(U, VDOj)^p \subset ext(DSi, VDOj)$*

Definition 4 Specialist Role. *A data source DSi plays a specialist role w.r.t. a condition p in a query on $VDOj$ iff most instances of $VDOj$ stored in DSi match p. $IsSpecialist(DSi, VDOj, p) \implies card(ext(DSi, VDOj)^p) \geq card(ext(DSi, VDOj)^{\neg p})$*

Definition 5 Container Role. *A data source DSi plays a container role w.r.t. a condition p in a query on $VDOj$ iff DSi contains at least one instance of $VDOj$ that matches p. $IsContainer(DSi, VDOj, p) \implies ext(U, VDOj)^p \cap ext(DSi, VDOj)^p \neq \varnothing$*

Data source roles are registered as extensional facts in the VO knowledge base and can be acquired using three approaches: (a) Manually (e.g. expert's or DBA's definition of knowledge), (b) Interpreting the execution of processes, (c) Automatically extracting it from sources of knowledge. In [20] we have presented strategies for the last two approaches.

Benefit model. In order to predict the benefit of a source DSi w.r.t. a predicate condition p the component uses Formula (1). This formula relates the role of the data source and the relative cardinality of the data source. The intention is to take into account the expected relevance of the source, given by the role factor, with the maximum number of instances that a data source may provide (in the best case). This combination of knowledge is necessary because sources that play a specialist role could be less important than sources that play container role if the number of instances the first group may provide is considerably lesser.

The *RoleFactor()* in Formula (2), returns a value between [0,1] indicating the most important role that a source may play in the query. *ContainerFactor,*

---

[1] This extension contains the object identifiers.

SpecialistFactor and *AuthorityFactor* are constants reflecting the importance of the role.

$$Benefit(DSi, VDOj, p) = RoleFactor(DSi, VDOj, p)*$$

$$\frac{card(ext(DSi, VDOj))}{\max\{card(ext(DSk, VDOj)), \forall\, DSk\ in\ U\ where\ RoleFactor(DSk, VDOj, p) > 0\}}$$

$$(1)$$

$$RoleFactor(DSi, VDOj, p) = \max((IsContainer(DSi, VDOj, p) * ContainerFactor),$$
$$(IsSpecialist(DSi, VDOj, p) * SpecialistFactor),$$
$$(IsAuthority(DSi, VDOj, p) * AuthorityFactor))$$

$$(2)$$

Although Formula (1) would be more accurate if the cardinality $card(ext(DSi, VDOj)^p)$ is used, the exact number of instances that satisfy a predicate is not available in VOs.

4.5   Joining Set Manager

To predict the set of sources that will not produce empty joins, the *Joining Set Manager* queries the VO knowledge base for obtaining the individuals of the composite VO units. This decision is made under the assumption that atomic VO units belonging to the same composite VO unit have more probability of containing the same group of instances. However, the creation of joining sets can use other type of rules to identify when two or more sources may share instances. For example, if two VO units are located in the same region, the probability that their data sources share instances may increase. Similarly, the fact that two VO units participate in the same VO process increases this probability.

Algorithm 1 uses the rule of composite units to create joining sets. The objective is to group together sources whose join will not produce an empty set and that are able to evaluate all the conditions of the query. It first [1] obtains the data sources of the composite units and creates one set for each of them. If there are redundant sets, the algorithm removes the set with fewer atomic units. Then, in [2], it validates the completeness of a set. A set is complete if the data sources it contains are able to evaluate all the conditions of the query. In [3] it extracts data sources from complete sets and [4] removes those incomplete sets whose data sources exist in the complete sets. If it is not the case, the algorithm gets the query conditions that are not already evaluated on each incomplete set [5] and completes these sets finding [6] the data source with higher role able to evaluate the missing conditions, from the complete sets.

The output of this component is the set of joining sets. A joining set must contain at least one data source to evaluate each condition of the query predicate. However, if more than one source may evaluate the same condition, it is not straightforward to establish which one of them should be selected to evaluate the condition. It would be necessary to determine which data source assignment would improve the final answer. The *Assignment Optimizer* component, described in Section 5, helps to make the right decision.

Algorithm 1 Joining Sets Creation

```
Input:  Q, CompositeUnits{ID1(DSi,..,DSj),...,IDm(DSk,...DSm)},
QRoles(DS1:role,...,DSn:role)
Output: JoiningSets{JS1(DSp,...,DSq),...,JSn(DSt,...,DSv)}
Begin
[1]initialSets{} = CreateInitialSets(CompositeUnits,QRoles)
   incompleteSets{} = {}       completeSets{} = {}
   ForAll (set in initialSets)
[2]    If (isComplete(set,Q)){ add(completeSets,set)}
       Else{ add(incompleteSets,set)}
   If (size(incompleteSets) > 0){
[3]  com{} = getDataSources(completeSets)//The data sources of complete sets
       ForAll (set in incompleteSets){
         inc{} = getDataSources(set)//The data sources of one incomplete set
         If ( contains(com, inc))  // com contains inc
[4]        remove(set, incompleteSets)
         Else{
[5]        conditions{} = getMissingEvaluation(set, Q)
           ForAll (cond in Conditions){
[6]          dataSource = getHighestRole(com,cond)
             addDataSource(set,dataSource)}
           add(completeSets,set)}
     } }
   Return(completeSets)
End
```

5   Optimizing Source Selection

The proposal is to see the problem of deciding which condition predicates are evaluated by which sources as an assignment problem [21] subject to resource constraints. We propose a mathematical model that receives as input the results of the benefit predictor component and the processing capacities of sources, if they are known. Although this model was created for source selection in large scale VOs, it can be used in any distributed query processing system during the planning phase.

5.1   Source Selection as an Assignment Problem

In this source selection problem, there are a number of predicate conditions to assign to a number of data sources. Each assignment brings some benefits in terms of response quality and consumes some resources of the sources. The objective is to maximize the benefits using the minimum of sources while satisfying the resource constraints. From the point of view of the predicate, one condition is assigned to one main source as *main assignment*, but this condition could also be evaluated as *secondary assignment* in parallel on other sources which have been selected by other conditions. The reason is that once a source is queried, it is better to evaluate there a maximum possible number of conditions, expecting to reduce the number of sources and the cost of transmitting instances that are not completely evaluated. Indeed, we have to deal with a bi-objective combinatorial optimization problem subject to semi-assignment and resource contraints. In practice, we choose to control the objective of minimizing the number of selected sources by converting it into a constraint. By default, the number of selected sources is limited to the number of conditions.

5.2   Mathematical Model

Given the set of data sources $I = \{1, ..., n\}$ and the set of predicate conditions $J = \{1, ..., m\}$, the input data are as follows:

-$Ben_{i,j}$: benefit of assigning condition j to source i, $\forall i \in I$, $\forall j \in J$;
-$MaxRes_i$: processing resource capacity of source i, $\forall i \in I$;
-$Res_{i,j}$: processing resources consumed in assigning condition j to source i, $\forall i \in I$, $\forall j \in J$;
-$MaxAssig_i$: maximal number of condition assignments for source i, $\forall i \in I$.
The decision variables are:
-$x_{i,j}$ are 0-1 variables that determine whether source i has (=1) or not (=0) been selected as a main source to evaluate the condition j, $\forall i \in I$, $\forall j \in J$.
-y_i are 0-1 variables that turn to 1 when the source i is selected, $\forall i \in I$.
-$assig_{i,j}$ are 0-1 variables that determine whether a condition j is assigned to source i (=1) or not (=0), $\forall i \in I$. These variables represent the final assignment of conditions to sources. The $x_{i,j}$ variables indicate the main assignments while the $assig_{i,j}$ variables indicate all the main and secondary assignments.

The mathematical program of the Source Selection Optimization Problem (SSOP) can be formulated as follows:

$$max \sum_{j=1}^{m} \sum_{i=1}^{n} Ben_{i,j} * (x_{i,j} + assig_{i,j}), \tag{3}$$

this objective function maximizes the benefit brought by the main and secondary source assignments of the conditions;
subject to

$$\sum_{i=1}^{n} x_{i,j} = 1, \forall j \in J, \tag{4}$$

constraints (4) ensure that any condition j is assigned to one main source i;

$$\sum_{i=1}^{n} y_i \leq k, \tag{5}$$

constraint (5) limits the total amount of queried sources to k, we take $k=m$ by default to start the optimization process;

$$\sum_{j=1}^{m} x_{i,j} \geq y_i, \forall i \in I, \tag{6}$$

these constraints (6) express that if a data source i is selected, there is at least one condition assigned to it, and reversely, if no condition is assigned to the data source i, this source is never selected;

$$x_{i,j} \leq assig_{i,j}, \forall i \in I, \forall j \in J, \tag{7}$$

$$assig_{i,j} \leq y_i, \forall i \in I, \forall j \in J, \tag{8}$$

these coupling constraints (7) and (8) indicate respectively that the main assignments should be in all main and secondary assignments as well, and that a source i is selected when at least one condition j is assigned to it;

$$\sum_{j=1}^{m} Res_{i,j} * assig_{i,j} \leq MaxRes_i, \forall i \in I, \tag{9}$$

$$\sum_{j=1}^{m} assig_{i,j} \leq MaxAssig_i, \forall i \in I, \tag{10}$$

these resource constraints (9) and (10) ensure that all the main and secondary assignments of conditions do not exceed neither the processing resource capacities nor the maximum number of possible assignments per source.

Notice that constraint (5) is somehow redundant with constraint (6) which prevents to select a source if no condition is assigned to it, i.e. the number of selected data sources is always less than or equal to the number of conditions m. But in practice, one could reduce k in constraint (5) to control the minimization of the number of selected sources.

The resolution of the model provides the selected sources in y_i variables and all the main and secondary assignments in $assig_{i,j}$ variables. The joining of the results provided by each condition are the instances required by the user. If the number of instances obtained from the joining sets are not enough to satisfy user requirements, the query planner will use the model to generate a new assignment with the remaining sources (i.e. those that were not in y_i).

5.3   Theoretical Complexity and Solving Methods

The Source Selection Optimization Problem (SSOP) is a knapsack-like problem. If we consider only the $assig_{i,j}$ variables and their related constraints, the problem is similar to a two-dimensional multiple knapsack problem which is NP-hard since it is a generalization of the knapsack problem. We have two types of resource constraints (9) and (10) for each data source (two dimensions for each knapsack) and multiple data sources (multiple knapsacks).

However, there are two differences. The first one is that one condition (item) is assigned to (i.e. evaluated by) one main data source (constraints (4)) and can be assigned also to several secondary data sources. The semi-assignment constraints (4) are only on the main assignment variables $x_{i,j}$. The second one is that we have to control the number k of the data sources selected via the variables y_i and constraint (5). This is similar to contraints of the uncapacitated facility location problem.

Indeed, if we consider the version of the SSOP in which we look only for the main assignements (more restricted than SSOP), denoted MSSOP, the MSSOP can be easily transformed into a multi-dimensional multiple knapsack problem which is NP-hard. Since the resolution of the MSSOP provides a feasible solution to the SSOP, the SSOP is also NP-hard.

To solve the SSOP, we have chosen to use in this study two well-known MIP solvers: CPLEX and GLPK, because the computational times obtained on the test data instances are acceptable. Naturally, heuristics and metaheuristics can be used to speedup the optimization if computational time constraint is tight and exact solutions are not required. In addition, heuristics and metaheuristics are also more suitable if multi-objective optimization is considered in this problem, i.e. altogether maximizing the benefit of the condition assignments to the data sources and minimizing the number of data sources selected.

6   Implementation and Validation

In order to evaluate the behavior of OptiSource and validate its improvement on the selection of the most relevant data sources, a prototype of OptiSource has been constructed and used to evaluate its precision and recall. This section presents the main results obtained during this evaluation. Section 6.1 presents the characteristics of the prototype. Section 6.2 details the experimental context. Section 6.3 gives the evaluation using different data contexts and levels of knowledge about sources. Finally, Section 6.4 compares our strategy to related works.

6.1   Prototype

OptiSource is part of a mediation system created for VOs. For evaluating the behavior of OptiSource we developed the components of OptiSource presented in Figure 7 as well as other components required for the query planning. The complete set of components of this prototype are presented in Figure 8. OptiSource is the *SourceSelector* component.

Components are written in Java. The knowledge base is implemented in OWL [19]. The optimization model is written in GNU MathProg modeling language and is processed by the GNU Linear Programming Kit (GLPK) [22]. We also validate our optimization model performance in CPLEX 10.2. For the implementation we use the Java Binding for GLPK called GLPK-Java [23]. Queries of VDOs are accepted in SPARQL [24] and are processed using Jena API and Pellet as the inference engine.

6.2   Experimental Context

To measure the extent to which the selection made by OptiSource is successful, it has been assessed whether OptiSource choose the data sources in order of relevance to the query. The evaluation has been conducted using evaluation

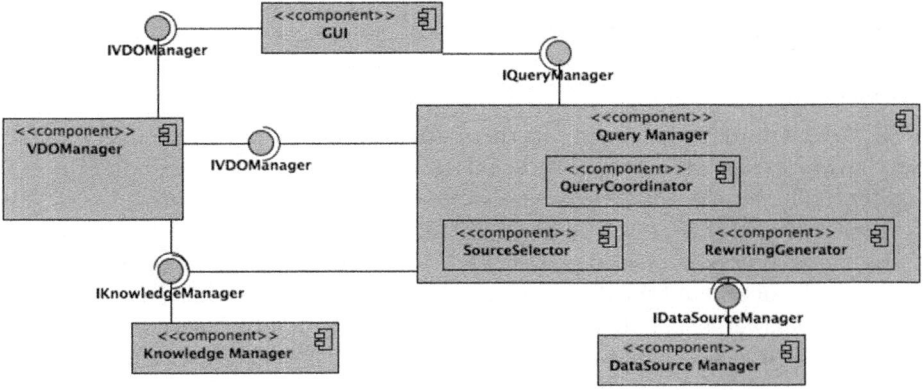

Fig. 8. Prototype Components

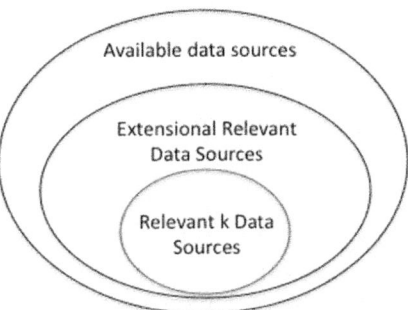

Fig. 9. Experimental Sets

measures that are typically applied in information retrieval systems, applying certain adjustments to the context. Figure 9 presents the sets of data sources that have been controlled during the experiments. The *Extensional Relevant Data Sources* are the sources that have at least one instance that matches the query predicate. The *Relevant k Data Sources* are the k data sources that have the higher contribution in terms of number of instances that match the query predicate, where k varies in each experiment.

The experiments are focused on measuring three metrics: *Extensional Precision*, that measures the extensional relevant data sources selected from the total set of data sources selected; *k Precision*, that measures the k relevant data sources selected from the total set of data sources selected; and the *k Recall*, that measures the first k data sources selected from the total set of k relevant data sources.

To obtain these metrics, we measure the following variables:

Rext: Data sources relevant extensionally for the query Q.
Rk: First k data sources relevant extensionally for the query Q.
A: Data sources selected by OptiSource.
Ak: First k data sources selected by OptiSource.

The formulas used to calculate each metric are the following:

Extensional Precision: $|A \cap Rext|/|A|$.

k Precision: $|Ak \cap Rk|/|Ak|$.

k Recall: $|Ak \cap Rk|/|Rk|$.

Experimental scenarios vary in three dimensions: the number of sources, the level of knowledge of sources and the relationship between the query and the data context. The number of sources can vary between 30 to 1000 data sources. The level of knowledge represents the precision and completeness of the knowledge facts included in the knowledge base w.r.t. the query. Changes on the relationship between the query and the data context cover the worst case of distribution where the query answer is distributed over more than 80% of data sources, and the best case where the answer is concentrated in less than 20% of data sources.

We define the knowledge base of a VO in the health domain that shares information about patients. Metadata are generated at different levels of the knowledge base using a developed generator of metadata in XML. This generator describes the intentional knowledge of data sources, the known roles of each data source and possible replication of instances with other data sources. For instance, a data source may declare: <class name ="Glaucoma" role="Container">, and another one can declare <class name="OperationontheEye" role="Container">.

6.3   Experimental Results

Due to space limitations we present only the most significant portion of the experiments executed. Figures 10 and 11 illustrate the impact of level of knowledge changes on the precision and recall of OptiSource. Low level means the knowledge base only contains the intentional knowledge of sources (e.g. can evaluate the diagnosis). The medium level means it also knows a group of roles of sources related to classes of the domain concept, but they are related to higher classes of the knowledge base. For instance, sources have related roles to *cancer* and queries ask for *kidney cancer*. High level indicates that roles are related to more specific classes (e.g. roles related to kidney cancer).

The experiments show that even if a high level of knowledge of the context is desired, OptiSource precision and recall have a good behavior with a medium level. This is due to the fact that a medium level of knowledge is enough to discard a large group of sources and to direct the queries to sources with higher probabilities of having matching instances. Additionally, and because of the use of an ontology to represent the knowledge base, even if the knowledge facts are not expressed at the same level of the query conditions, the inference engine infers knowledge facts that enhance the estimation of the benefit.

Figure 12 presents the impact of the relationship between the query and the data context on the extensional precision and the k precision. It also compares the results with a strategy oriented to capabilities. Figure 12 illustrates the behavior of OptiSource when the instances are concentrated in less than 20% of the available data sources. In such cases OptiSource exceeds by more than 50 % the extensional precision and the k precision of the capability oriented strategy. This is the result of source reduction performed by OptiSource using the roles

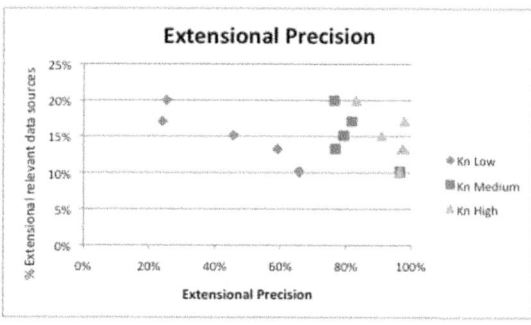

Fig. 10. Extensional precision evaluation changing the level of knowledge

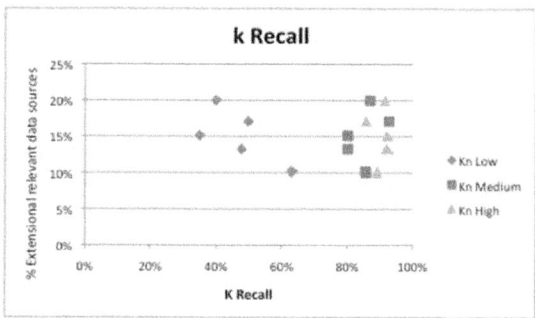

Fig. 11. k Recall evaluation changing the level of knowledge

and the cardinality of data sources. OptiSource precision is stable and always higher than the strategies that only use intentional knowledge and capabilities.

Tests also show that for data contexts with fewer data sources, the improvement in precision according to the level of knowledge is not significant. This validates our assumption that in small data contexts great efforts in query planning for reducing queried sources are not necessary. Still, the improvement observed when there is a large number of sources led us to conclude that OptiSource is especially useful under these circumstances.

Finally, the experiments show that in the worst case (low level of knowledge) OptiSource selects the same set of sources that traditional strategies of query rewriting select because it only considers the intentional views of sources. Other experimental results related to the precision and recall of OptiSource are presented in [9].

6.4   Comparison with Related Works

Although OptiSource cannot be directly compared with any of the proposals in the literature, we evaluated the implications of using two available strategies (QPIAD [6] and Navigational Paths [14]) in VO contexts. We also adapted the proposal iDrips [5] since it is the closest to OptiSource in its proposed intention.

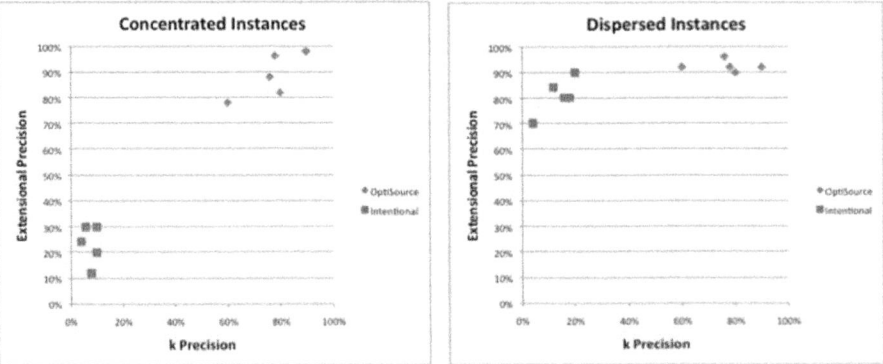

Fig. 12. Extensional precision and k precision evaluation changing the relationship between the query and the data context

The strategy of rewriting of QPIAD [6] allows the query processor to return certain and "possible" certain answers from databases. The first type are answers that match all the query predicate; the second type are the answers that probably match the query, but have one condition that cannot be evaluated because the required value is missing (*null*). QPIAD generates and uses detailed statistics on source's quality to reformulate a query according to the characteristics of each source that is able to evaluate the predicate. Even though the principle of QPIAD will be very attractive for use in VOs whose sources have quality problems, the strategy could not scale up when the number of sources is large. To prove this we will consider a VO with 20 sources, which are able to evaluate a query Q in terms of the schema. We supposed that in average each source provides 500 certain answers. For each source 50 query reformulations are required to obtain the possible answers. This means that 1020 (*50*20+20*) queries are executed including the original Q to obtain certain answers from each source.

Another strategy connected to VO contexts is the proposal to improve the selection of navigational paths between biological sources [14]. It assumes the availability of a graph describing the objects contained in the sources and their relationships. The objective is to find the path with best benefit and lower cost given an initial and a final object class. Applying this proposal to our problem we found that it is possible to organize sources in a graph. Each node could contain sources with equivalent intentional fragments of VDO, and the relationships between sources that contain the missing fragments of VDO could be the links between different nodes. For instance, if sources 1 and 2 know the *Demographic Data* and *Medical Act* classes of *Patient* they will be in the same node, and will be related to nodes that contain sources that know the *Affiliation* class. Using this graph it will be possible to select the most "profitable" sources for a query; however, it is not clear how the benefit of each path is obtained and how the overlapping between each path can be calculated. This lack could be combined with our strategy using the roles of data sources in each path to compute its potential benefit. Thus, the prediction model of OptiSource can be used to compute the ratios between paths used in this proposal.

Finally, we compared OptiSource with the iDrips algorithm. This comparison is focused on measuring the recall in terms of relevant instances obtained vs. the number of sources queried. We did not measure precision because iDrips and OptiSource always return correct instances. In order to make comparable the measurements of recall in OptiSource and iDrips, we evaluate in each iteration of iDrips the number of relevant instances until it finds the complete set of relevant instances. We also measure the number of queried sources in each iteration. In the case of OptiSource, we measure the relevant instances obtained after the execution of subqueries of the joining sets.

We used the plan coverage as the utility measure for iDrips and three formulas of utility that led us to analyze the different levels of knowledge of the VO context. Given a plan $P = \{DS1, ..., DSn\}$ where DSi are the sources that will be queried on the plan and a query $Q = \{p(p_1, ..., p_m)\}$ with m predicate conditions, the three formulas to compute plan coverage are as follows:

$$Cov(P)_{T1} = min(ext(DSi, VDOj)) \; where \; DSi \in P, \tag{11}$$

this formula assumes that the number of instances of the VDO sources contain is known;

$$Cov(P)_{T2} = min(overlap(DSi, DSk)) \; where \; DSi \; and DSk \in P, \tag{12}$$

this formula assumes the same knowledge of (11) and the level of overlapping between data sources;

$$Cov(P)_{T3} = \frac{ext(DSi, VDOj)^p - (ext(DSk, VDOj)^p ... ext(DSl, VDOj)^p)}{ext(U, VDOj)^p} \tag{13}$$

where $DSi, DSk \; and \; DSl \in P \; and \; DSk...DSl$ have been already queried in a previous plan. Formula (13) assumes the knowledge of (11) and (12). It also requires knowing the number of instances that satisfy the query predicate in each source (or in each plan) and from these instances which ones can also be provided by another plan. Even though this information will be difficult and expensive to obtain in a VO context, we used it because the creators of iDrips assumed having it when they designed the algorithm.

The first group of experiments are run using a set of synthetic sources that are extensionally and intentionally overlapped, and where a few number of sources contain a large number of relevant instances; even though instances are distributed largely along the VO sources. In other words, in these contexts there are sources specialized in one or more of the query conditions. Consequently, querying them is enough to obtain almost all the relevant instances. iDrips is evaluated using the aforementioned utility measures. Figure 13 illustrates the differences on recall between iDrips and OptiSource. The y axis shows the percentage of queried sources from the total number of relevant data sources. On the other hand, the x axis illustrates the percentage of instances obtained.

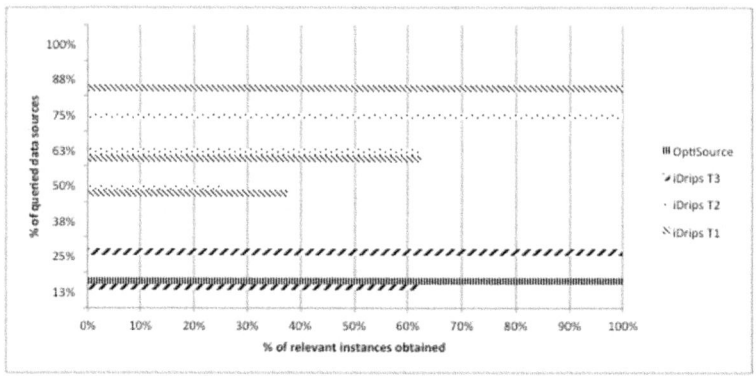

Fig. 13. OptiSource vs. iDrips

The results demonstrate the effectiveness of OptiSource in this type of contexts where it only queried 13% of relevant sources to obtain 100% of relevant instances. On the contrary, iDrips required to query 25% of relevant sources to obtain the 100% of relevant instances in the best case (using Formula (13)). Figure 13 shows that the number of sources queried by OptiSource is considerably lower than those queried using (11) and (12) of the plan coverage. Although this could be inferred considering the difference on the level of knowledge, it allows us to conclude that solutions for ranking queries cannot be applied to VOs without using a powerful utility function able to represent source relationships and source probability of having query relevant instances. Nonetheless, even with strong utility functions, the absence of replication knowledge prevents discarding sources that do not contribute with new relevant instances. This lack affects iDrips when we use Formula (13).

We have also run experiments in a context where synthetic data sources are overlapped intentionally but not extensionally, and a small percentage of sources are specialized in one or more of the conditions of the predicate. As expected, the behavior of both strategies are similar because it is necessary to contact almost all the sources related intentionally to the query.

7   Analyzing the Impact of Prediction Accuracy

The most important input of the optimization model is the benefit of using a data source to evaluate a query condition. As it has been illustrated, the value of the benefit is predicted taking into account the role of the data source and its relative size. If the prediction is well done the optimization will be very accurate. However, if there is not certainty about the validity of the prediction, the model will probably produce inaccurate results. Thus, this section presents first the methodology used to evaluate the impact of the benefit value in the final assignment, then it presents the obtained results and an analysis over them.

7.1 Analysis Methodology

Sensitivity analysis of optimal solutions is a well-known task that allows to evaluate the impact of changes in input data parameters of linear programming models. Although there are well-known methods to evaluate the sensitivity of the optimization model results, due to the high degeneracy of the assignment problem, these traditional methods are impractical in our case [25]. Additionally, even though there are some proposals to determine the sensitivity range of parameters in the assignment problem [25], they could not be applied to our model because they assumed the use of the classical assignment problem, and although our model is close to the assignment problem, it differs from the classical one by its resource constraints. Besides, they assumed the unimodular property of constraint matrix, which is not checked by the matrix of our model. Consequently, in order to evaluate the impact of the model, we set up an experimental study framework, which allows us to measure the impact of the benefit values. The idea behind is to measure changes on the assignment when the value of the benefit values are known under uncertainty.

To ensure that the experiments are significant, the model is applied to different sets of data that change the number of data sources (25,50,100,200,400,800) and the number of conditions (5,10,20); at the end 18 sets are produced. For each one of the sets, a benefit matrix is created ($Ben_{i,j}$ in Section 5) with real values: this matrix is called the initial matrix. Using this initial matrix the proposed model is executed using GLPK. The assignment obtained from the execution of the model for each set is considered as the base result for each set.

The initial matrix is then modified in order to simulate the uncertainty of the value of each benefit. The precision of the value is reduced randomly, i.e. the new values of $Ben_{i,j}$ can vary from $0.33 \times Ben_{i,j}$ to $0.66 \times Ben_{i,j}$. The intention of these reductions is to change the value of the benefit in a similar way as it is reduced when the role of a data source or the size of a data source is not known precisely. These new matrices produce secondary results that are compared with the base results. For each initial matrix, 50 new matrices are generated for each set.

7.2 Experimental Results

Figure 14 illustrates a summary of the experiment results. The x axis represents the percentage of assignments that change in an experiment. The y axis represents the percentage of experiments for which it has been obtained a change in the assignment. Each line represents the number of data sources in the experiment. For example, with 100 data sources, 11% of experiments did not have any change on the assignment when the benefits are changed. Similarly, with 800 data sources 22% of experiments had 60% of changes on the assignment when the benefit changes. As it can be noticed, changes variability is directly related to the number of data sources. This result has been expected because the higher the number of data sources is, the greater the probability of finding a data source with a better profit value that forces to change the assignment is. The study also demonstrated that most part of the experiments had changes in

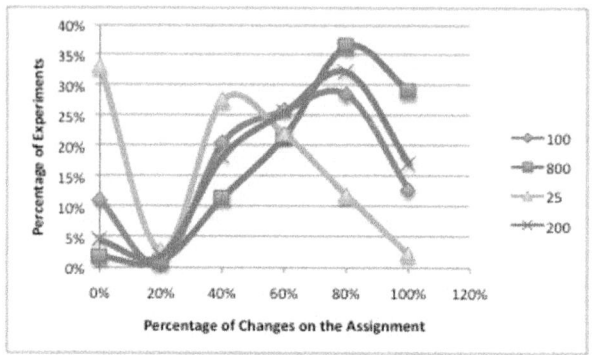

Fig. 14. General Impact of the Benefit

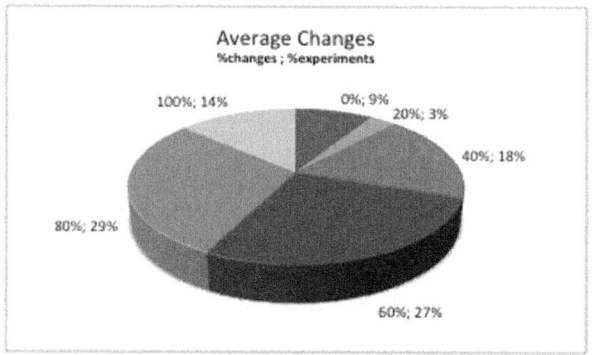

Fig. 15. Average of changes in all the experiment sets

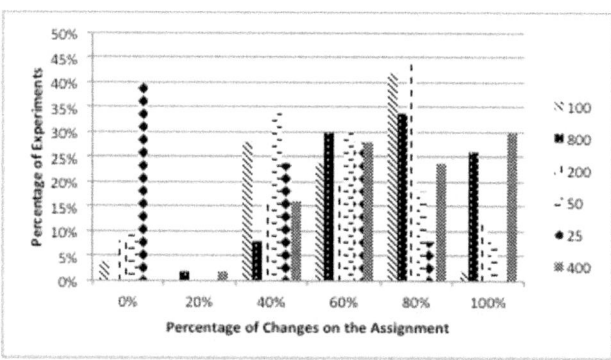

Fig. 16. Impact of the Benefit in the Assignment - 15%

the assignment between 40% and 70%. Besides, the tests show that 25% of the assignments had changes of 60% independently on the number of data sources. Figure 15 shows the average of experiments for each set of changes. For example,

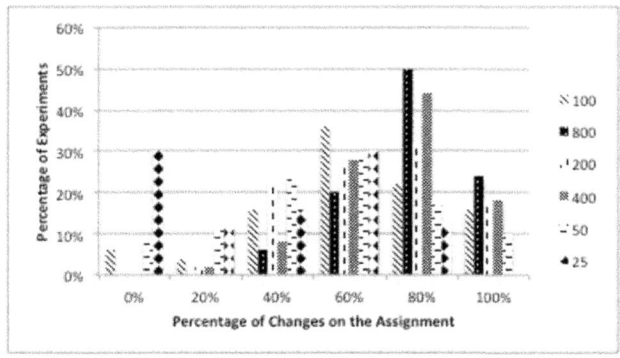

Fig. 17. Impact of the benefit in the assignment - 50%

27% of experiments had changes near 60%, and 18% of experiments had changes near 40%.

Figure 16 and Figure 17 present the percentage of changes when the benefit was changed a maximum of 15% or a maximum of 50%, respectively. As in the general case, most part of experiments had assignment changes between 40% to 70%, independently on the percentage of changes on the benefit value.

Results on sensibility show that changes on the values of the benefit, as expected, affects the final response. However, a percentage from 20% to 60% of the assignments remain the same, even if the values of the benefit changes 50% from the base value. This fact, demonstrates a good level of robustness of our optimization model, which does not require a precise prediction in order to make an assignment near to the optimal.

8   Conclusions and Future Work

Virtual Organizations including a large number of autonomous participants that collaborate to share knowledge are very common in web contexts. The number and autonomy of the participants prevent a global understanding of the data characteristics and their relationships. This leads to complex data contexts where query processing is expensive. One of the main factors for that is the difficulty to effectively select the sources to access for a query evaluation. As current strategies do not focus on contexts with numerous heterogenous overlapping sources their use in large scale VOs will lead to the generation of redundant queries to unnecessary sources.

In order to tackle this difficulty, we have modeled the source selection problem as a complex decision problem using combinatorial optimization techniques. This leaded us to combine several original contributions including knowledge representation and a mathematical formulation of the source selection problem.

This paper presented the main choices and OptiSource, the proposed source selection strategy. It is based on the prediction of the benefit of using a source during query evaluation according to its individual and group relevance. The combinatorial optimization model is used to select the most relevant sources to be used to evaluate each query condition. The main choices and contributions are the following:

- **The context** is represented with a generic ontology that can be specialized according to the problem domain. This ontology facilitates the description of a VO. We proposed a general representation including VOUnits, VOResources, VODomainConcepts and their relationships. This enhances the knowledge of the mediator and facilitates its evolution if necessary.

- We proposed the use of **roles** for defining relationships between sources, concepts and queries. Roles represent a hierarchical relationship between sources and concepts that enable the source prioritization during the source selection phase. The roles presented in this paper are inspired by a health domain. Nevertheless, our proposal is generic and the roles can be customized according to the application domain. Roles contribute to the estimation of the individual contribution of a source to evaluate a query predicate.

- **OptiSource adopts a combinatorial optimization model** that establishes which sources evaluate each condition of the query. This model uses the knowledge about data sources and their relationships. The discovery process of such knowledge is based on Virtual Organization description (such as composite VO Units), their business processes, geographical location of participants among others. Additional relationships between participants can be expressed by experts or deduced from historical query process.

- **A prototype of OptiSource** has been developed in Java and used GLPK-Java. The knowledge base is implemented in OWL. The optimization model is written in GNU MathProg modeling language and is processed with the GNU Linear Programming Kit (GLPK). CPLEX 10.2 has also been used to validate the combinatorial optimization model. Queries of VDOs are accepted in SPARQL and are evaluated using Jena API and Pellet as the inference engine.

- Numerous **experiments** have been conducted, in particular to evaluate the precision and the robustness of our proposal. Experiments have demonstrated that OptiSource is effective to reduce the contact of unnecessary sources when data contexts involve extensional and intentional overlapping. They also showed that OptiSource is an evolving system whose precision grows according to the degree of knowledge of sources, but does not require a high level of knowledge to have good levels of precision and recall. Sensibility analysis of OptiSource w.r.t. the benefit estimation demonstrates a good level of robustness of the optimization model, which does not require a precise prediction in order to make an assignment near to the optimal.

This work opens important research issues. Among them, the extension of the work of extracting and using Virtual Organization knowledge to improve data management. The use of data mining techniques on logs of business processes and query executions is promising. The combination of such knowledge with statistics should also be further developed. Future research also involves the use of our proposal in other contexts leading probably to other source relationships and data source roles.

To finish, probably the most important research perspective we consider, is the further use of combinatorial optimization techniques for the optimization of large scale information systems, such as multi-objective optimization and metaheuristic solving methods to enable more complex modeling of the source selection optimization problem. Our experience is very positive and we intend to pursuit this approach.

References

1. Foster, I., Kesselman, C., Tuecke, S.: The anatomy of the grid: Enabling scalable virtual organizations. International Journal of High Performance Computing Applications 15, 200–222 (2001)
2. NEESGrid: Nees consortium (2008), http://neesgrid.ncsa.uiuc.edu/
3. BIRN: Bioinformatics research network (2008), http://www.loni.ucla.edu/birn/
4. Quiané-Ruiz, J.-A., Lamarre, P., Valduriez, P.: Sqlb: A query allocation framework for autonomous consumers and providers. In: VLDB, pp. 974–985 (2007)
5. Doan, A., Halevy, A.Y.: Efficiently ordering query plans for data integration. In: ICDE 2002, p. 393. IEEE Computer Society, Washington, DC, USA (2002)
6. Wolf, G., Khatri, H., Chokshi, B., Fan, J., Chen, Y., Kambhampati, S.: Query processing over incomplete autonomous databases. In: VLDB, pp. 651–662 (2007)
7. Huebsch, R., Hellerstein, J.M., Lanham, N., Loo, B.T., Shenker, S., Stoica, I.: Querying the internet with pier. In: VLDB, Berlin, Germany, pp. 321–332 (2003)
8. Pottinger, R., Halevy, A.Y.: Minicon: A scalable algorithm for answering queries using views. VLDB Journal. 10(2-3), 182–198 (2001)
9. Pomares, A., Roncancio, C., Cung, V.-D., Abásolo, J., Villamil, M.-d.-P.: Source selection in large scale data contexts: An optimization approach. In: Bringas, P.G., Hameurlain, A., Quirchmayr, G. (eds.) DEXA 2010. LNCS, vol. 6261, pp. 46–61. Springer, Heidelberg (2010)
10. Levy, A.Y., Rajaraman, A., Ordille, J.J.: Querying heterogeneous information sources using source descriptions. In: VLDB, pp. 251–262 (1996)
11. Garcia-Molina, H., Papakonstantinou, Y., Quass, D., Rajaraman, A., Sagiv, Y., Ullman, J.D., Vassalos, V., Widom, J.: The tsimmis approach to mediation: Data models and languages. Journal of Intelligent Information Systems 8, 117–132 (1997)
12. Tomasic, A., Raschid, L., Valduriez, P.: Scaling access to heterogeneous data sources with DISCO. Knowledge and Data Engineering 10, 808–823 (1998)
13. Yerneni, R.: Mediated Query Processing Over Autonomous Data Sources. PhD thesis, Stanford University, Stanford, CA (2001)
14. Bleiholder, J., Khuller, S., Naumann, F., Raschid, L., Wu, Y.: Query planning in the presence of overlapping sources. In: Ioannidis, Y., Scholl, M.H., Schmidt, J.W., Matthes, F., Hatzopoulos, M., Böhm, K., Kemper, A., Grust, T., Böhm, C. (eds.) EDBT 2006. LNCS, vol. 3896, pp. 811–828. Springer, Heidelberg (2006)

15. Naumann, F., Freytag, J.C., Leser, U.: Completeness of integrated information sources. Information Systems Journal, Special issue: Data Quality in Cooperative Information Systems 29, 583–615 (2004)
16. Tatarinov, I., Ives, Z., Madhavan, J., Halevy, A., Suciu, D., Dalvi, N., Dong, X(L.), Kadiyska, Y., Miklau, G., Mork, P.: The piazza peer data management project. SIGMOD Rec. 32(3), 47–52 (2003)
17. Nejdl, W., Wolf, B., Qu, C., Decker, S., Sintek, M., Naeve, A., Nilsson, M., Palmér, M., Risch, T.: Edutella: a p2p networking infrastructure based on rdf. In: WWW 2002, pp. 604–615. ACM, New York (2002)
18. Adjiman, P., Goasdoué, F., Rousset, M.-C.: Somerdfs in the semantic web. Journal on Data Semantics 8, 158–181 (2007)
19. Horrocks, I.: Owl: A description logic based ontology language. In: Principles and Practice of Constraint Programming, pp. 5–8 (2005)
20. Pomares, A., Roncancio, C., Abasolo, J., del Pilar Villamil, M.: Knowledge based query processing. In: ICEIS. Lecture Notes in Business Information Processing, vol. 24, pp. 208–219. Springer, Heidelberg (2009)
21. Hillier, F.S., Lieberman, G.J.: Introduction to Operations Research, 8th edn. McGraw-Hill, New York (2005)
22. Makhorin, A.: Gnu project, gnu linear programming kit (2009), http://www.gnu.org/software/glpk/
23. Makhorin, A.: Gnu project, glpk for java (2009), http://glpk-java.sourceforge.net/
24. Eric Prud, A.S.: Sparql query language for rdf (2007), http://www.w3.org/tr/rdf-sparql-query/
25. Lin, C.-J., Wen, U.-P.: Sensitivity analysis of the optimal assignment. European Journal of Operational Research 149(1), 35–46 (2003)

On the Benefits of Transparent Compression for Cost-Effective Cloud Data Storage

Bogdan Nicolae

INRIA Saclay, Île-de-France
bogdan.nicolae@inria.fr

Abstract. Infrastructure-as-a-Service (IaaS) cloud computing has rev-
olutionized the way we think of acquiring computational resources: it
allows users to deploy virtual machines (VMs) at large scale and pay
only for the resources that were actually used throughout the runtime of
the VMs. This new model raises new challenges in the design and development of IaaS middleware: excessive storage costs associated with both
user data and VM images might make the cloud less attractive, especially
for users that need to manipulate huge data sets and a large number of
VM images. Storage costs result not only from storage space utilization,
but also from bandwidth consumption: in typical deployments, a large
number of data transfers between the VMs and the persistent storage are
performed, all under high performance requirements. This paper evaluates the trade-off resulting from transparently applying data compression
to conserve storage space and bandwidth at the cost of slight computational overhead. We aim at reducing the storage space and bandwidth
needs with minimal impact on data access performance. Our solution
builds on BlobSeer, a distributed data management service specifically
designed to sustain a high throughput for concurrent accesses to huge
data sequences that are distributed at large scale. Extensive experiments
demonstrate that our approach achieves large reductions (at least 40%)
of bandwidth and storage space utilization, while still attaining high
performance levels that even surpass the original (no compression) performance levels in several data-intensive scenarios.

1 Introduction

The emerging cloud computing model [28, 2, 1] is gaining serious interest from
both industry [31] and academia [16,35,23] for its proposal to view the computation as a utility rather than a capital investment. According to this model, users
do not buy and maintain their own hardware, nor have to deal with complex
large-scale application deployments and configurations, but rather rent such resources as a service, paying only for the resources their computation has used
throughout its lifetime.

There are several ways to abstract resources, with the most popular being
Infrastructure-as-a-Service (IaaS). In this context, users rent raw computational
resources as virtual machines that they can use to run their own custom applications, paying only for the computational power, network traffic and storage

A. Hameurlain, J. Küng, and R. Wagner (Eds.): TLDKS III , LNCS 6790, pp. 167–184, 2011.

space used by their virtual environment. This is highly attractive for users that cannot afford the hardware to run large-scale, distributed applications or simply need flexible solutions to scale to the size of their problem, which might grow or shrink in time (e.g. use external cloud resources to complement their local resource base [14]).

However, as the scale and variety of data increases at a fast rate [5], the cost of processing and maintaining data remotely on the cloud becomes prohibitively expensive. This cost is the consequence of excessive utilization of two types of resources: *storage space* and *bandwidth*.

Obviously, a large amount of storage space is consumed by the data itself, while a large amount of bandwidth is consumed by the need to access this data. Furthermore, in order to achieve scalable data processing performance, users often employ data intensive paradigms (such as MapReduce [3] or Dryad [10]) that generate massively parallel accesses to the data, and have additional *high aggregated throughput* requirements.

Not so obvious is the additional storage space and bandwidth utilization overhead introduced by operating the virtual machine images that host the user application. A common patten on IaaS clouds is the need to deploy a large number of VMs on many nodes of a data-center at the same time, starting from a set of VM images previously stored in a persistent fashion. Once the application is running, a similar challenge applies to snapshotting the deployment: many VM images that were locally modified need to be concurrently transferred to stable storage with the purpose of capturing the VM state for later use. Both these patterns lead to a high storage space and bandwidth utilization. Furthermore, they have the same high aggregated throughput requirement, which is crucial in order to minimize the overhead associated with virtual machine management.

Therefore, there is a need to optimize three parameters simultaneously: (1) conserve storage space; (2) conserve bandwidth; and (3) deliver a high data-access throughput under heavy access concurrency.

This paper focuses on evaluating the benefits of applying data compression *transparently* on the cloud, with the purpose of achieving a good trade-off for the optimization requirements mentioned above. Our contributions can be summarized as follows:

- We propose a generic sampling-based compression layer that dynamically adapts to heterogeneous data in order to deal with the highly concurrent access patterns generated by the deployment and execution of data-intensive applications. This contribution extends our previous proposal presented in [18]. In particular, we introduce a more generic compression layer that extends the applicability of our proposal to the management of virtual machine images (in addition to the management of application data) and show how to integrate it in the cloud architecture.
- We propose an implementation of the compression layer on top of *Blob-Seer* [17, 20], a versioning-oriented distributed storage system specifically designed to deliver high throughputs under heavy access concurrency.

– We perform extensive experimentations on the Grid5000 testbed [12] that demonstrate the benefits of our approach. In particular, in addition to the experiments presented in [18], we highlight the benefits of our approach for virtual machine image storage.

2   Our Approach

In this section we present an adaptive, transparent compression layer that aims at reducing the space and bandwidth requirements of cloud storage with minimal impact on I/O throughput when under heavy access concurrency.

2.1   General Considerations

Several important factors that relate to the data composition and access patterns need to be taken into consideration when designing a compression layer for the cloud. We enumerate these factors below:

Transparency. In a situation where the user is running the application on private-owned hardware and has direct control over the resources, compression can be explicitly managed at application level. This approach is often used in practice because it has the advantage of enabling the user to tailor compression to the specific needs of the application. However, on clouds, explicit compression management at application level is not always feasible. For example, many cloud providers offer data-intensive computing platforms (such as Elastic MapReduce [32] from Amazon) directly as a service to their customers. These platforms are based on paradigms (such as MapReduce [3]) that abstract data access, forcing the application to be written according to a particular schema which makes explicit compression management difficult.

Furthermore, besides application data, users customize and store virtual machine images on the cloud that are then used to deploy and run their application. However, users are not allowed to directly control the deployment process of virtual machine images and therefore cannot apply custom compression techniques on the images.

For these two reasons, it is important to handle compression *transparently* and offer it as an extra feature to the users, potentially reducing their storage and bandwidth costs with minimal impact on quality-of-service.

Random access support. Most compression algorithms are designed to work with data streams: new data is fed as input to the algorithm, which in turn tries to shrink it by using a dynamic dictionary to replace long patterns that were previously encountered with shorter ones. This is a process that destroys the original layout of the data: a subsequence of the original uncompressed data stream that starts at an arbitrary offset cannot be easily obtained from the compressed data stream without decompressing the whole data.

However, random access to data is very common on clouds. For example, a virtual machine typically does not access the whole contents of the underlying virtual machine image during its execution: only the parts that are needed to boot the virtual machine and run the user application are accessed. Furthermore, application data is typically organized in huge data objects that comprise many small KB-sized records, since it is unfeasible to manage billions of small, separate data objects explicitly [7]. The data is then processed in a distributed fashion: different parts of the same huge data object are read and written concurrently by the virtual machines. Again, this translates to a highly concurrent random read and write access pattern to the data objects.

Therefore, it is important to design a compression layer that is able to overcome the limitations of stream-based compression algorithms and introduce support for *efficient random access* to the underlying data.

Heterogeneity of data. First of all, compression is obviously only useful as long as it shrinks the space required to store data. However, on clouds, the data that needs to be stored is highly heterogeneous in nature.

Application data is mostly in unstructured form. It either consists of *text* (e.g., huge collections of documents, web pages and logs [25]) or *multimedia* (e.g., images, video and sound [26]). While text data is known to be highly compressible, multimedia data is virtually not compressible and in most cases trying to apply any compression method on it actually increases the required storage space and generates unnecessary computational overhead.

Virtual machine images typically represent the contents of the virtual disks attached to the virtual machine. Some parts are claimed by the file system and hold executables, archived logs, etc. These parts are typically not compressible. Other parts are not used by the file system (zero-sequences of bytes) or hold logs, configuration files, etc., which makes them highly compressible.

Thus, the compression layer needs to dynamically adapt to the type of data stored on the cloud, dealing efficiently with both compressible and incompressible data.

Computational overhead. Compression and decompression invariably leads to a computational overhead that diminishes the availability of compute cores for effective application computations. Therefore, this overhead must be taken into account when designing a high-performance compression layer. With modern high-speed networking interfaces, high compression rates might become available only at significant expense of computation time. Since the user is not paying only for storage space and bandwidth, but for the CPU utilization as well, choosing the right trade-off is often difficult.

Memory overhead. Deploying and running virtual machines takes up large amounts of main memory from the nodes of the cloud that host them. Given this context, main memory is a precious resource that has to be carefully managed. It is therefore crucial to design a compression layer that minimizes the extra main memory required for compression and decompression.

2.2   Design Principles

In order to deal with the issues presented above, we propose the following set of design principles:

Data striping. A straight-forward way to apply compression is to fully compress the data before sending it remotely in case of a write operation, respectively to wait for the compressed data to arrive and then decompress it in case of a read operation. However, this approach has a major disadvantage: the compression/decompression does not run in parallel with the data transfer, potentially wasting computational power that is idle during the transfer. For this reason, we propose the use of data striping: the piece of data is split into chunks and each chunk is compressed independently. This way, in the case of a write, a successfully compressed chunk can be sent before all other chunks have finished compressing, while in the case of a read, a fully received chunk can be decompressed before all other chunks have been successfully received. Thus, data striping enables overlapping of compression with data transfers, increasing the overall achieved throughput.

At the same time, data striping deals with the random access limitation of many compression algorithms. By compressing each chunk individually, reads and writes and random offsets involve only the chunks that cover the requested range delimited by offset and size. If the data is split at fine granularity, the overhead of random access becomes negligible. However, a chunk size that is too small may limit the potential of compression because fewer repeating patterns appear in smaller chunks. Therefore, it is important to find the right trade-off when choosing the chunk size.

Sampling of chunks. Since the system needs to adapt to both compressible and incompressible data, there is a need to determine efficiently when to compress a chunk and when to leave it in its original form. Obviously, attempting to compress the whole chunk and evaluating the result generates unacceptably high overhead. Therefore, we need a way to predict whether it is useful to apply compression or not. For this reason, we propose to *sample* each chunk, i.e. pick a small random part of the chunk and apply compression on it. Under the assumption that the obtained compression ratio predicts the compression ratio that would have been obtained by compressing the whole chunk itself, the chunk will be compressed only if the compression ratio of the small piece of random data is satisfactory.

The risk of wrong predictions is very low, because the chunk size is much smaller compared to the size of the whole data object, making it likely that the data of the small random part is of the same nature as the data of the whole chunk. Even when a prediction is wrong, given the small size of the chunk, the impact is minimal: either a chunk remains uncompressed when it should have been compressed, which has no negative impact on performance and a minimal impact on saved storage space, or, a chunk is compressed when it shouldn't, which has a minimal computational overhead and no negative impact on storage space (because the result of the compression is discarded and the chunk is stored in its original uncompressed form).

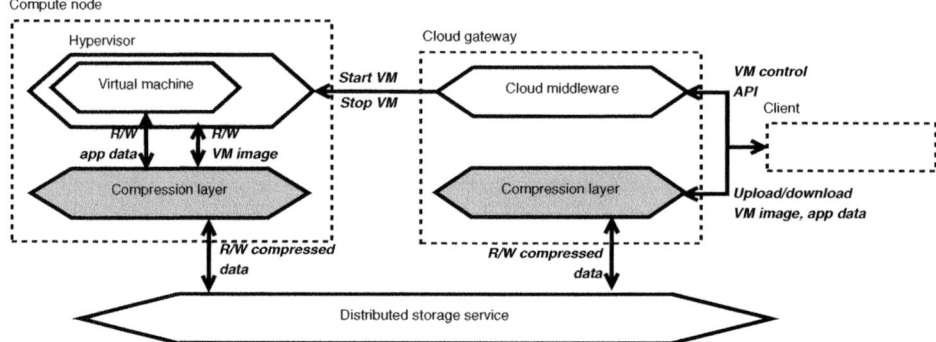

Fig. 1. Cloud architecture that integrates our proposal (dark background)

Configurable compression algorithm. Dealing with the computation and memory overhead of compressing and decompressing data is a matter of choosing the right algorithm. A large set of compression algorithms have been proposed in the literature that trade off compression ratio for computation and memory overhead. However, since the compression ratio relates directly to storage space and bandwidth costs, the user should be allowed to configure the algorithm in order to be able to fine-tune this trade-off according to the needs.

2.3 Architecture

Starting from the design principles presented above, we propose a compression layer that integrates in the cloud as shown in Figure 1. The typical elements found in the cloud are illustrated with a light background, while the compression layer is highlighted by a darker background.

The following actors are present:

- **Cloud middleware:** is responsible to manage the physical resources on the cloud: it schedules where new virtual machines are instantiated, it keeps track of consumed resources for each user, it enforces policies, etc. The cloud middleware exposes a control API that enables users to perform a wide range of management tasks: VM deployment and termination, monitoring, etc.
- **Distributed storage service:** is responsible to organize and store the data on the cloud. It acts as a data sharing service that facilitates transparent access to the data within given quality-of-service guarantees (performance, data availability, etc.) that are established by the cloud provider in the service level agreement.
- **Cloud client:** it uses the control API of the cloud middleware in order to interact with the cloud. It also accesses the data storage service in order to manipulate virtual machine images and application data.
- **Hypervisor:** is the virtualization middleware that leverages the physical resources of the compute nodes to present a virtual operating platform for the virtual machines. In this role, it emulates a virtual file system that is

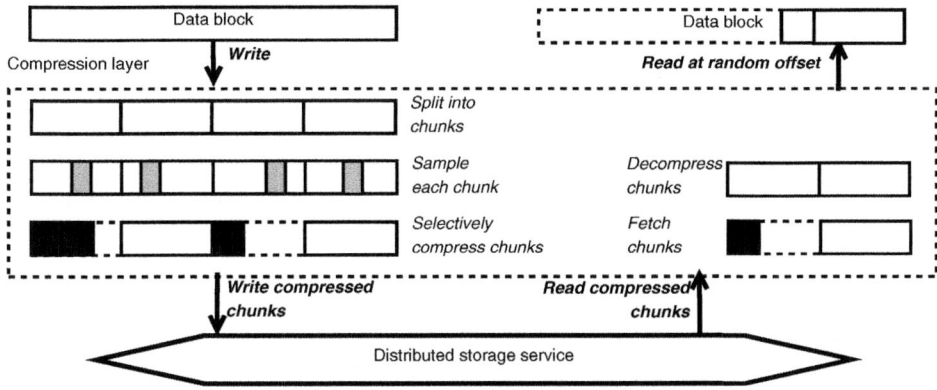

Fig. 2. Zoom on the compression layer

backed up by a virtual machine image, which is a regular file that is accessible from the compute node's host file system.

- **Virtual machine:** represents the virtual environment in which the guest operating system and user applications are running. Virtual machines can communicate with each other and share application data through the distributed storage service.
- **Compression layer:** traps all data accesses (both application data and virtual machine images) and treats them according to the principles presented in Section 2.2. It runs both on the compute nodes and on the cloud gateways, mediating the interactions of the clients, hypervisors and virtual machines with the distributed storage service.

Figure 2 zooms on the compression layer, which is responsible to trap all read and write accesses to the distributed storage service.

In case a write operation is performed, after the data is split into chunks, a small random sample of each chunk is compressed in order to probe whether the chunk is compressible or not. If the achieved compression ratio is higher than a predefined threshold, then the whole chunk is compressed and the result is written to the distributed storage service. If the achieved compression ratio is lower than the threshold, then the chunk is written directly to the distributed storage service without any modification.

In case a read operation is performed, first all chunks that cover the requested range (delimited by offset and size) are determined. These chunks are then fetched from the storage service and decompressed if they were stored in compressed fashion. The uncompressed contents of each chunk is then placed at its relative offset in the local buffer supplied by the application. The read operation succeeds when all chunks have been successfully processed this way, filling the local buffer.

In both cases, the compression layer processes the chunks in a highly parallel fashion, potentially taking advantage of multi-core architectures. This enables overlapping of remote transfers to and from the storage service with the compression and decompression to high degree, minimizing the latency of read and write operations due to compression overhead.

Furthermore, since the compression layer is running on the client-side of the storage service (i.e. directly on the compute nodes and cloud gateways), the burden of compression and decompression is not falling on the storage service itself, which greatly enhances performance under specific access patterns, such as the case when the same chunk is accessed concurrently from multiple nodes.

3   Implementation

In this section we show how to efficiently implement our proposal such that it both achieves the design principles introduced in Section 2.2 and is easy to integrate in the cloud as shown in Section 2.3.

We have chosen to leverage *BlobSeer*, presented in Section 3.1, as the distributed storage service on top of which to implement our approach. This choice was motivated by two factors. First, BlobSeer implements out-of-the-box transparent data striping of large objects and fine-grain access to them, which enables easy implementation of our approach as it eliminates the need for explicit chunk management. Second, BlobSeer offers support for high throughput under concurrency, which enables efficient parallel access the chunks and therefore is crucial to achieving our high performance objective.

3.1   BlobSeer

This section introduces BlobSeer, a distributed data storage service designed to deal with the needs of data-intensive applications: *scalable aggregation of storage space* from the participating nodes with minimal overhead, support to store *huge data objects*, *efficient fine-grain access* to data subsets and ability to sustain a *high throughput under heavy access concurrency*.

Data is abstracted in BlobSeer as long sequences of bytes called BLOBs (Binary Large OBject). These BLOBs are manipulated through a simple access interface that enables creating a blob, reading/writing a range of *size* bytes from/to the BLOB starting at a specified *offset* and appending a sequence of *size* bytes to the BLOB. This access interface is designed to support versioning explicitly: each time a write or append is performed by the client, a new snapshot of the blob is generated rather than overwriting any existing data (but physically stored is only the difference). This snapshot is labeled with an incremental version and the client is allowed to read from any past snapshot of the BLOB by specifying its version.

Architecture. BlobSeer consists of a series of distributed communicating processes. Each BLOB is split into chunks that are distributed among *data providers*. *Clients* read, write and append data to/from BLOBs. Metadata is associated to each BLOB and stores information about the chunk composition of the BLOB and where each chunk is stored, facilitating access to any range of any existing snapshot of the BLOB. As data volumes are huge, metadata grows to significant sizes and as such is stored and managed by the *metadata providers* in a decentralized fashion. A *version manager* is responsible to assign versions to snapshots and ensure high-performance concurrency control. Finally, a *provider manager* is responsible to employ a chunk allocation strategy, which decides what chunks are stored on which data providers, when writes and appends are issued by the clients. A *load-balancing* strategy is favored by the provider manager in such way as to ensure an even distribution of chunks among providers.

Key features. BlobSeer relies on *data striping, distributed metadata management* and *versioning-based concurrency control* to avoid data-access synchronization and to distribute the I/O workload at large-scale both for data and metadata. This is crucial for achieving a high aggregated throughput under concurrency, as demonstrated by our previous work [20, 22, 19, 15].

3.2   Integration with BlobSeer

Since BlobSeer implicitly performs data striping whenever a data block is written into it, we implemented the compression layer directly on top of the client-side networking layer of BlobSeer, which is responsible for remote communication with the data providers.

Instead of directly reading and writing the chunks from/to BlobSeer, the compression layer acts as a filter that performs the sampling and compresses the chunks when the requested operation is a write, respectively decompresses the chunks if the requested operation is a read. Careful consideration was given to keep the memory footprint to a minimum, relying in the case of incompressible chunks on *zero-copy* techniques, which eliminate the need to copy chunks from one memory region to another when they remain unchanged and are passed to the networking layer.

The compression layer was designed to be highly configurable, such that any compression algorithm can be easily plugged in. For the purpose of this paper we adopted two popular choices: Lempel-Ziv-Oberhumer(LZO) [34], based on the work presented in [30], which focuses on minimizing the memory and computation overhead, and BZIP2 [33], a free and open-source standard compression algorithm, based on several layers of compression techniques stacked on top of each other.

The versioning-based BLOB access API exposed by BlobSeer can be leveraged at application level directly, as it was carefully designed to enable efficient access to user data under heavy access concurrency. However, in order to leverage the compression layer for efficient storage of virtual machine images, we relied on our previous work presented in [21]: a dedicated virtual file system on build on top of Blob-Seer, specifically optimized to efficiently handle two recurring virtual machine image access patterns on the cloud: multi-deployment and multi-snapshotting.

4   Experimental Evaluation

In this section we evaluate the benefits of our approach by conducting a series of large-scale experiments that target the access patterns typically found on the clouds. In particular, we focus on two settings: (1) read and write access patterns as generated by data-intensive applications (Sections 4.2 and 4.3) and (2) multi-deployments of virtual machines (Section 4.4).

The motivation behind choosing the access patterns for the first setting is the fact that data-intensive computing paradigms are gaining increasing popularity as a solution to cope with growing data sizes. For example, MapReduce [3] has been hailed as a revolutionary new platform for large-scale, massively parallel data access [24]. Applications based on such paradigms continuously acquire massive datasets while performing (in parallel) large-scale computations over these datasets, generating highly concurrent read and write access patterns to user data. We therefore argue that experimenting with such access patterns is a good predictor of the potential benefits achievable in practice.

The motivation behind the choice for the second setting is the fact that multi-deployments are the most frequent pattern encountered on the clouds. Users typically need to deploy a distributed application that requires a large number of virtual machine instances, however, for practical reasons it is difficult to manually customize a separate virtual machine image for each instance. Therefore, users build a single virtual machine image (or a small initial set) that they upload to cloud storage and then use it as a template to initialize a large number of instances from it, which ultimately leads to the multi-deployment pattern.

In both settings we are interested in evaluating both the access performance of our approach (throughput and execution time), as well as the reductions in network traffic and storage space when compared to the original BlobSeer implementation that does not implement a compression layer.

4.1   Experimental Setup

We performed our experiments on the Grid'5000 [12] testbed, a highly configurable and controllable experimental Grid platform gathering 9 sites in France. We used nodes belonging to two sites of Grid'5000 for our experiments: Rennes (122 nodes) and Lille (45 nodes). Each node is outfitted with dual-core or quadcore x86_64 CPUs (capable of hardware virtualization support) and have at least 4 GB of RAM. We measured raw buffered reads from the hard drives at an average of about 60MB/s, using the *hdparm* utility. Internode bandwidth is 1 Gbit/s (we measured 117.5 MB/s for TCP end-to-end sockets with MTU of 1500 B) and latency is 0.1 ms.

4.2   Concurrent Writes of Application Data

This scenario corresponds to a typical user data acquisition phase, in which the user application, consisting of many distributed processing elements, gathers application data concurrently from multiple sources (e.g. web crawling or log parsing) and stores it in a huge BLOB for later processing. The scenario generates a write-intensive access pattern in which data is appended concurrently to the same BLOB. We aim at evaluating our approach under such heavy access concurrency circumstances, both in the case when the data to be processed is compressible and in the case when it is not.

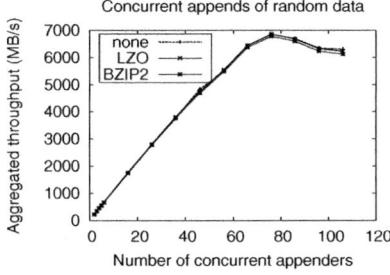

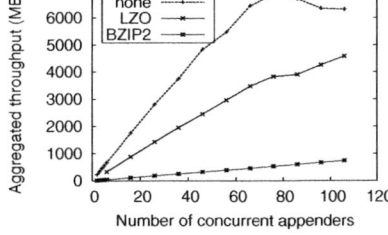

(a) Writing incompressible random data: high aggregated throughput is ensured by negligible sampling overhead.

(b) Writing compressible text data: high aggregated throughput when using LZO.

Fig. 3. Impact of our approach on aggregated throughput under heavy concurrency. In both cases concurrent clients append each 512 MB of data which is transparently split into 64MB chunks.

In order to perform this evaluation, we use 122 nodes of the Rennes site and deploy BlobSeer on them as follows: 110 data providers are deployed on different nodes, with an additional 10 dedicated nodes reserved to deploy the metadata providers. The version manager and provider manager are deployed on dedicated nodes as well.

Both in the case of compressible and incompressible data, we measure the aggregated throughput achieved when N concurrent clients append 512 MB of data in chunks of 64MB to the same BLOB. Each of the clients runs in its own virtual machine that is co-deployed with a data provider on the same node. Each data provider is configured to use a cache of 512MB, which is deducted from the total available main memory for the client.

In the first case that corresponds to compressible data, we use the text of books available online. Each client builds the sequence of 512MB by assembling text from those books. In the second case, the sequence of 512MB is simply randomly generated, since random data is the worst case scenario for any compression algorithm.

We perform experiments in each of the cases using our implementation (for both LZO and BZIP2) and compare it to the reference BlobSeer implementation that does not integrate the compression layer. Each experiment is repeated three times for reliability and the results are averaged. The sample size used to decide whether to compress the chunk or not is fixed at 64KB.

The obtained results are represented in Figure 3. The curves corresponding to random data (Figure 3(a)) are very close, clearly indicating that the *impact of sampling is negligible*, both for LZO and BZIP2. On the other hand, when using compressible text data (Figure 3(a)), the aggregated throughput in the case of LZO, although scaling, is significantly lower than the total aggregated throughput achieved when not compressing data. With less than 1 GB/s maximal aggregated throughput, performance levels in the case of BZIP2 are rather poor.

When transferring uncompressed data, an interesting effect is noticeable: past 80 concurrent appenders, the aggregated throughput does not increase but rather slightly decreases and then stabilizes. This effect is caused by the fact that concurrent transfers of such large amounts of data saturate the physical bandwidth limit of the system, which limits the achievable scalability.

With respect to storage space, gains from storing text data in compressed form are represented in Figure 4(b). With a consistent gain of about 40% of the original size, LZO compression is highly attractive. Although not measured explicitly, the same gain can be inferred for bandwidth utilization too. In the case of BZIP2, the gain reaches well over 60%, which makes up for the poor throughput.

4.3   Concurrent Reads of Application Data

This scenario is complementary to the previous scenario and corresponds to a highly concurrent data processing phase in which the user application concurrently reads and processes different parts of the same BLOB in a distributed fashion.

Since our approach stores incompressible data in its original form, there is no difference between reading incompressible data using our approach and reading the data without any compression layer enabled. For this reason, we evaluate the impact of our approach for compressible data only. Assuming text data was written in compressed form as presented in the previous section, we aim at evaluating the total aggregated throughput that can be sustained by our approach when reading the data back.

We use the same 122 nodes of the Rennes site for our experiments and keep the same deployment settings: 110 data providers are deployed on different nodes, while each of the N clients is co-deployed with a data provider on the same node. One version manager, one provider manager and 10 metadata providers are each deployed on a dedicated node. We measure the aggregated throughput achieved when N concurrent clients read 512 MB of data stored in compressed chunks, each corresponding to 64MB worth of uncompressed data. Each client is configured to read a different region of the BLOB, such that no two clients access the same chunk concurrently, which is the typical case encountered in the data processing phase.

As with the previous setting, we perform three experiments and average the results. All clients of the same experiment read from the region of the BLOB generated by the corresponding append experiment, i.e. the first read experiment reads the data generated by the first append experiment, etc. This ensures that no requested data was previously cached and forces a "cold" run each time.

The results are represented in Figure 4(a). In the case of uncompressed data transfers, the aggregated throughput stabilizes at about 7 GB/s, because large data sizes are transferred by each client, which saturates the networking infrastructure. On the other hand, using LZO compression brings substantial read throughput improvements: the transfer of smaller compressed chunks combined with the fast decompression speed on the client side contribute to a steady

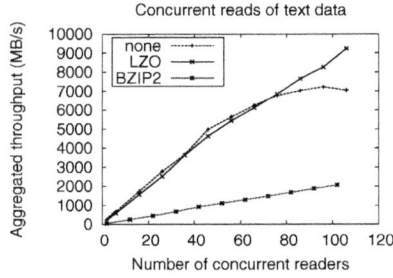
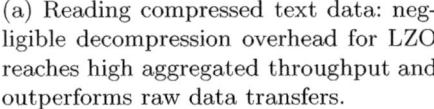
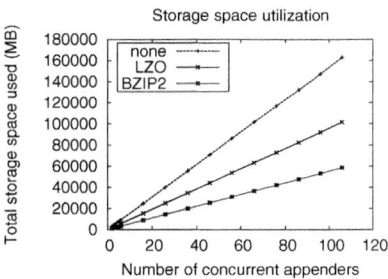

(a) Reading compressed text data: negligible decompression overhead for LZO reaches high aggregated throughput and outperforms raw data transfers.

(b) Counting the totals: BZIP2 saves more than 60% of storage space and bandwidth utilization. LZO reaches 40%.

Fig. 4. Impact of our approach on compressible text data: concurrent clients read 512MB of compressed data saved in chunks of 64MB (left); total bandwidth and storage space conserved (right).

increase in aggregated throughput that reaches well over 9 GB/s, which surpasses the bandwidth limits of the networking infrastructure and therefore would have been impossible to reach if data was not compressed. With a maximal aggregated throughput of about 2 GB/s, BZIP2 performs much better at reading data, but the results obtained are still much lower than compared to LZO.

4.4   Concurrent Accesses to Virtual Machine Images

Finally, we perform a series of experiments that evaluate the benefits of using compression for virtual machine image storage. In this context, the clients are not the user application that runs inside the virtual machines, but rather the hypervisors that execute the virtual machines and need to access the underlying virtual machine images stored in the cloud.

We assume the following typical scenario: the user has customized a virtual machine image and has uploaded it on the cloud, with the purpose of using it as a template for deploying a large number of virtual machines simultaneously. Once the user has asked the cloud middleware to perform this multi-deployment, each hypervisor instantiates its corresponding virtual machine, which in turn boots the guest operating system and runs the application.

In order to implement this scenario in our experiments, we use 45 nodes of the Lille cluster and deploy BlobSeer on it in the following fashion: 40 data providers, 3 metadata providers, one version manager and one provider manager. Each process is deployed on a dedicated node.

Next, we store a 2 GB large virtual machine image (a Debian Sid Linux distribution) in BlobSeer in three configurations: (1) no compression layer (the original BlobSeer implementation); (2) our compression layer with LZO compression, and (3) our compression layer with BZIP2 compression. In all three configurations, the image is split into 2 MB chunks.

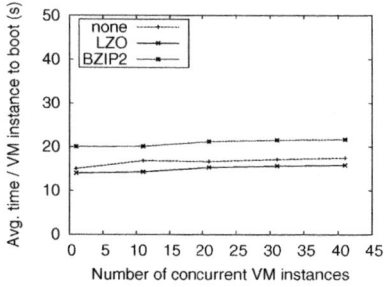
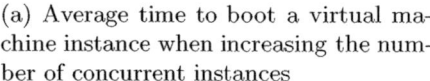

(a) Average time to boot a virtual machine instance when increasing the number of concurrent instances

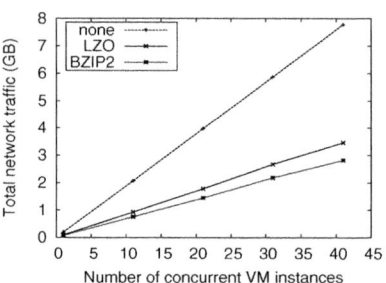

(b) Total network traffic generated by the boot process for all instances

Fig. 5. Performance results when concurrently deploying a large number of virtual machines from the same virtual machine image that is stored in compressed fashion using our approach.

We use the same nodes where data providers are deployed as compute nodes. Each of these nodes runs KVM 0.12.4 as the hypervisor. The experiment consists in performing a series of increasing multi-deployments from the virtual machine image that was previously stored in BlobSeer, for all three configurations. We perform two types of measurements: the average time taken by each instance to fully boot the guest operating system and the total amount of network traffic that was generated by the hypervisors as a result of reading the parts of the virtual machine image that were accessed during the boot process.

The obtained results are shown in Figure 5. The average time to boot an instance is depicted in Figure 5(b). As can be observed, in all three configurations the average time to boot an instance increases slightly as more instances are booted concurrently. This effect is due to the fact that more concurrent read accesses to the same data put more pressure on BlobSeer, which lowers throughput slightly. Nevertheless, the curves are almost constant and demonstrate the high scalability of our approach, in all three configurations. The fast decompression time of LZO combined with the lower amounts of data transfers give a constant boost of performance to our approach of more than 20% over the original Blob-Seer implementation. On the other hand, due to higher decompression times, BZIP2 performs 30% worse than the original implementation.

Figure 5(b) illustrates the total network traffic incurred in all three configurations. As expected, the growth is linear and is directly proportional to the amount of data that was read by the hypervisors from the virtual machine image. When the image is stored in uncompressed fashion, the total network traffic is close to 8 GB for 40 instances. Using LZO compression lowers the total network traffic by more than 50%, at well below 4 GB. Further reductions, reaching more than 60%, are observed using BZIP2 compression. In this case the total network traffic is about 3 GB, bringing the highest reduction in bandwidth cost of all three configurations.

5   Related Work

Data compression is highly popular in widely used data-intensive application frameworks such as Hadoop [8]. In this context, compression is not managed transparently at the level of the storage layer (Hadoop Distributed File System [9]), but rather explicitly at the application level. Besides introducing complexity related to seeking in compressed streams, this approach is also not aware of the I/O performed by the storage layer in the background, which limits the choice of optimizations that would otherwise be possible, if the schedule of the I/O operations was known.

Adaptive compression techniques that apply data compression transparently have been proposed in the literature before.

In [11], an algorithm for transferring large datasets in wide area networks is proposed, that automatically adapts the compression effort to currently available network and processor resources in order to improve communication speed. A similar goal is targeted by ACE [13] (Adaptive Compression Environment), which automatically applies on-the-fly compression at the network stack directly to improve network transfer performance. Other work such as [6] applies on-the-fly compression at higher level, targeting an improve in response time of web-services by compressing the exchanged XML messages. Although these approaches conserve network bandwidth and improve transfer speed under the right circumstances, the focus is end-to-end transfers, rather than total aggregated throughput. Moreover, compression is applied in-transit only, meaning data is not stored remotely in a compressed fashion and therefore requests for the same data generate new compression-decompression cycles over and over again.

Methods to improve the middleware-based exchange of information in interactive or collaborative distributed applications have been proposed in [29]. The proposal combines methods that continuously monitor current network and processor resources and assess compression effectiveness, deciding on the most suitable compression technique. While this approach works well in heterogeneous environments with different link speeds and CPU processing power, in clouds resources are rather uniform and typically feature high-speed links, which shifts the focus towards quickly deciding if to apply compression at all, and, when it is the case, applying fast compression techniques.

Several existing proposals define custom virtual machine image file formats, such as QCOW2 [4] and MIF [27]. These formats are able to hold the image in compressed fashion. However, unlike our approach, compression and decompression is not transparent and must be handled at the level of the hypervisor directly. While this has the advantage of enabling the hypervisor to optimize the data layout in order to achieve better compression rates, our approach has an important benefit in that it is non-intrusive: it handles compression independently of the hypervisor. This greatly improves the portability of the images, compensating for the lack of image format standardization.

6   Conclusions

As cloud computing gains in popularity and data volumes grow continuously to huge sizes, an important challenge of data storage on the cloud is the *conservation of storage space and bandwidth*, as both resources are expensive at large scale and can incur high costs for the end user.

This paper evaluates the benefits of applying *transparent* compression for data storage services running on the cloud, with the purpose of reducing costs associated to storage space and bandwidth, but without sacrificing data access performance for doing so. Unlike work proposed so far that focuses on end-to-end data transfer optimizations, we target to achieve a *high total aggregated throughput*, which is a more relevant metric in the context of clouds.

Our approach integrates with the storage service and adapts to heterogeneous data dynamically, by *sampling small portions of data on-the fly* in order to avoid compression when it is not beneficial. We *overlap compression and decompression with I/O*, by splitting the data into chunks and taking advantage of multi-core architectures, therefore minimizing the impact of compression on total throughput. Finally, we enable *configurable compression algorithm selection*, which enables the user to fine-tune the trade-off between computation time costs and storage and bandwidth costs.

We show a negligible impact on aggregated throughput when using our approach for incompressible data thanks to negligible sampling overhead and a high aggregated throughput both for reading and writing compressible data that brings massive storage space and bandwidth saves ranging between 40% and 60%. Our approach works well both for storing application data, as well as for storing virtual machine images. Using our approach with fast compression and decompression algorithms, such as LZO, higher application data access throughputs and faster virtual machine multi-deployments can be achieved under concurrency, all with the added benefits of lower storage space and bandwidth utilization.

Thanks to our encouraging results, we plan to explore in future work more adaptability approaches that are suitable in the context of data-intensive applications and virtual machine image storage. In particular, so far we used fixed chunk sizes and compression algorithms. An interesting future direction would be to dynamically select the chunk size and the compression algorithm for each chunk individually such as to reduce storage space and bandwidth consumption even further.

Acknowledgments

The experiments presented in this paper were carried out using the Grid'5000/ ALADDIN-G5K experimental testbed, an initiative from the French Ministry of Research through the ACI GRID incentive action, INRIA, CNRS and RENATER and other contributing partners (see http://www.grid5000.fr/ for details).

References

1. Armbrust, M., Fox, A., Griffith, R., Joseph, A., Katz, R., Konwinski, A., Lee, G., Patterson, D., Rabkin, A., Stoica, I., Zaharia, M.: A view of cloud computing. Commun. ACM 53, 50–58 (2010)
2. Buyya, R., Yeo, C.S., Venugopal, S., Broberg, J., Brandic, I.: Cloud computing and emerging IT platforms: Vision, hype, and reality for delivering computing as the 5th utility. Future Gener. Comput. Syst. 25(6), 599–616 (2009)
3. Dean, J., Ghemawat, S.: MapReduce: simplified data processing on large clusters. Communications of the ACM 51(1), 107–113 (2008)
4. Gagné, M.: Cooking with Linux: still searching for the ultimate linux distro? Linux J. 2007(161), 9 (2007)
5. Gantz, J.F., Chute, C., Manfrediz, A., Minton, S., Reinsel, D., Schlichting, W., Toncheva, A.: The diverse and exploding digital universe: An updated forecast of worldwide information growth through 2011. IDC (2007)
6. Ghandeharizadeh, S., Papadopoulos, C., Pol, P., Zhou, R.: Nam: a network adaptable middleware to enhance response time of web services. In: MASCOTS 2003: Proceedings of the 11th IEEE/ACM International Symposium on Modeling, Analysis and Simulation of Computer Telecommunications Systems, pp. 136–145 (12-15 2003)
7. Ghemawat, S., Gobioff, H., Leung, S.T.: The Google file system. SIGOPS - Operating Systems Review 37(5), 29–43 (2003)
8. The Apache Hadoop Project, http://www.hadoop.org
9. HDFS. The Hadoop Distributed File System, http://hadoop.apache.org/common/docs/r0.20.1/hdfs_design.html
10. Isard, M., Budiu, M., Yu, Y., Birrell, A., Fetterly, D.: Dryad: distributed data-parallel programs from sequential building blocks. SIGOPS Oper. Syst. Rev. 41(3), 59–72 (2007)
11. Jeannot, E., Knutsson, B., Björkman, M.: Adaptive online data compression. In: HPDC 2002: Proceedings of the 11th IEEE International Symposium on High Performance Distributed Computing, p. 379. IEEE Computer Society, Washington, DC, USA (2002)
12. Jégou, Y., Lantéri, S., Leduc, J., Noredine, M., Mornet, G., Namyst, R., Primet, P., Quetier, B., Richard, O., Talbi, E.G., Iréa, T.: Grid'5000: a large scale and highly reconfigurable experimental grid testbed. International Journal of High Performance Computing Applications 20(4), 481–494 (2006)
13. Krintz, C., Sucu, S.: Adaptive on-the-fly compression. IEEE Trans. Parallel Distrib. Syst. 17(1), 15–24 (2006)
14. Marshall, P., Keahey, K., Freeman, T.: Elastic site: Using clouds to elastically extend site resources. In: CCGRID 2010: Proceedings of the 10th IEEE/ACM International Conference on Cluster, Cloud and Grid Computing, pp. 43–52. IEEE Computer Society, Washington, DC, USA (2010)
15. Montes, J., Nicolae, B., Antoniu, G., Sánchez, A., Pérez, M.: Using Global Behavior Modeling to Improve QoS in Cloud Data Storage Services. In: CloudCom 2010: Proceedings of the 2nd IEEE International Conference on Cloud Computing Technology and Science, Indianapolis, USA, pp. 304–311 (2010)
16. Moreno-Vozmediano, R., Montero, R.S., Llorente, I.M.: Elastic management of cluster-based services in the cloud. In: ACDC 2009: Proceedings of the 1st Workshop on Automated Control for Datacenters and Clouds, pp. 19–24. ACM, New York (2009)

17. Nicolae, B.: BlobSeer: Towards Efficient Data Storage Management for Large-Scale, Distributed Systems. Ph.D. thesis, University of Rennes 1 (November 2010)
18. Nicolae, B.: High throughput data-compression for cloud storage. In: Globe 2010: Proceedings of the 3rd International Conference on Data Management in Grid and P2P Systems, Bilbao, Spain, pp. 1–12 (2010)
19. Nicolae, B., Antoniu, G., Bougé, L.: Enabling high data throughput in desktop grids through decentralized data and metadata management: The BlobSeer approach. In: Sips, H., Epema, D., Lin, H.-X. (eds.) Euro-Par 2009. LNCS, vol. 5704, pp. 404–416. Springer, Heidelberg (2009)
20. Nicolae, B., Antoniu, G., Bougé, L., Moise, D., Carpen-Amarie, A.: Blobseer: Next-generation data management for large scale infrastructures. J. Parallel Distrib. Comput. 71, 169–184 (2011)
21. Nicolae, B., Bresnahan, J., Keahey, K., Antoniu, G.: Going Back and Forth: Efficient Multi-Deployment and Multi-Snapshotting on Clouds. In: HPDC 2011: Proceedings of the 20th International ACM Symposium on High-Performance Parallel and Distributed Computing, San José, CA, USA (2011)
22. Nicolae, B., Moise, D., Antoniu, G., Bougé, L., Dorier, M.: Blobseer: Bringing high throughput under heavy concurrency to hadoop map/reduce applications. In: IPDPS 2010: Proc. 24th IEEE International Parallel and Distributed Processing Symposium, Atlanta, USA, pp. 1–12 (2010)
23. Nurmi, D., Wolski, R., Grzegorczyk, C., Obertelli, G., Soman, S., Youseff, L., Zagorodnov, D.: The Eucalyptus open-source cloud-computing system. In: CC-GRID 2009: Proceedings of the 9th IEEE/ACM International Conference on Cluster, Cloud and Grid Computing, pp. 124–131. IEEE Computer Society, Los Alamitos (2009)
24. Patterson, D.A.: Technical perspective: the data center is the computer. Commun. ACM 51(1), 105–105 (2008)
25. Pavlo, A., Paulson, E., Rasin, A., Abadi, D.J., DeWitt, D.J., Madden, S., Stonebraker, M.: A comparison of approaches to large-scale data analysis. In: SIGMOD 2009: Proceedings of the 35th SIGMOD International Conference on Management of Data, pp. 165–178. ACM, New York (2009)
26. Raghuveer, A., Jindal, M., Mokbel, M.F., Debnath, B., Du, D.: Towards efficient search on unstructured data: an intelligent-storage approach. In: CIKM 2007: Proceedings of the 16th ACM Conference on Information and Knowledge Management, pp. 951–954. ACM, New York (2007)
27. Reimer, D., Thomas, A., Ammons, G., Mummert, T., Alpern, B., Bala, V.: Opening black boxes: using semantic information to combat virtual machine image sprawl. In: VEE 2008: Proceedings of the 4th ACM SIGPLAN/SIGOPS International Conference on Virtual Execution Environments, pp. 111–120. ACM, New York (2008)
28. Vaquero, L.M., Rodero-Merino, L., Caceres, J., Lindner, M.: A break in the clouds: towards a cloud definition. SIGCOMM Comput. Commun. Rev. 39(1), 50–55 (2009)
29. Wiseman, Y., Schwan, K., Widener, P.: Efficient end to end data exchange using configurable compression. SIGOPS Oper. Syst. Rev. 39(3), 4–23 (2005)
30. Ziv, J., Lempel, A.: A universal algorithm for sequential data compression. IEEE Transactions on Information Theory 23, 337–343 (1977)
31. Amazon Elastic Compute Cloud (EC2), http://aws.amazon.com/ec2/
32. Amazon Elastic Map Reduce, http://aws.amazon.com/elasticmapreduce/
33. BZIP2, http://bzip.org
34. Lempel-Ziv-Oberhumer, http://www.oberhumer.com/opensource/lzo
35. The Nimbus project, http://www.nimbusproject.org

A Mobile Web Service Middleware and Its Performance Study

Jingyu Zhang[1], Shiping Chen[2],
Yongzhong Lu[3], and David Levy[4]

[1] Faculty of Engineering and Information Technologies,
The University of Sydney, Australia
jing.zhang@sydney.edu.au
[2] Information Engineering Laboratory, CSIRO ICT Centre, Australia
Shiping.Chen@csiro.au
[3] School of Software Engineering, Huazhong University of Science and Technology,
Wuhan, China
YongzhongLu@gmail.com
[4] Faculty of Engineering and Information Technologies,
The University of Sydney, Australia
David.Levy@sydney.edu.au

Abstract. Web services have been widely accepted as a platform independent services-oriented technology. Meanwhile, ubiquitous technologies are becoming popular in a variety of domain applications. In particular, hosting web services from mobile devices became a way to extend knowledge exchange and share. This paper presents our design and implementation of a mobile web service (MWS) middleware motivated by a mobile application that assists observers in the surveillance and diagnosis of animal diseases in the field. We also present a performance study of hosting web services on mobile devices by evaluating the MWS. Based on our observations, a performance model is used to predict the performance of a class of MWS-based applications.

Keywords: mobile web service, middleware, soap attachment, performance model, performance prediction.

1 Introduction

As mobile devices have become so popular in our lives, the demand for mobility is extending beyond traditional telecommunications. For example, email service and web browsing are widely running on handheld devices. As the computational power of mobile devices is increasing, the "move to mobile" [1] is one of the central themes in the services domain. Mobile devices trend to be a pervasive interface for a variety of applications, such as electronic mobile wallets [2], learning terminals [3], and clients of cloud computing [4].

The majority of the above mobile applications play a client-side role. However, in some circumstances, a mobile device is required to work as a services provider, for example, a group of people working collaboratively in a rural area

A. Hameurlain, J. Küng, and R. Wagner (Eds.): TLDKS III , LNCS 6790, pp. 185–207, 2011.

where one cannot access internet or mobile base stations. In animal pathology, epidemiological surveillance and diagnosis has gradually become a top priority around the world. An effective epidemiological surveillance network for high-risk animal diseases is based on general rules. In northern pastoral regions of Australia, extensive beef cattle production is characterised by large herds, large distances between farms and infrequent use of veterinary services. Most livestock observation opportunities occur in remote field locations. These factors combine to make disease surveillance difficult in this sector [5-7]. As paddocks are very remote from the nearest veterinarian service stations, the awareness of the cattle diseases is much more difficult. Furthermore, there is no public network coverage in paddocks, which means that farmers do not have the opportunity to access veterinarian services via phone or internet.

In [3] and [8], we presented a mobile epidemiological surveillance and diagnosis system for in-field training of emerging infectious diseases. The mobile bovine syndromic surveillance system (mBOSSS) is a stand-alone application developed to support epidemiological surveillance within remote northern pastoral regions of Australia. This project aimed to study the observation methods of diseased livestock from users (not necessarily veterinarians) in the farm industry. With mBOSSS, the users describe the symptoms of the observed disease. These reported symptoms are compared to a large database of symptoms and diseases to identify the potential cause of the disease and to guide further investigation. Ease of use and convenience of data capture are key aspects that will determine the success and quality of this system. However, when the database is too large to be hosted in a mobile phone, the data must be distributed on multiple mobile devices shared across several observers. Therefore, as mBOSSS cannot collaborate with peers, the data source capacity limits mBOSSS from providing a complete syndromic surveillance service on mobile devices.

Web service technologies have been widely accepted as a solution to interoperability of applications. A web service is an approach to a variety of mobile applications deployed onto heterogeneous mobile platforms in term of mobile device hardware and Operating System.

There are a few middleware packages available for mobile web services [9-12] and most of them only support a simple SOAP messaging. In our in-field diagnosis application, a mobile service is required to provide multimedia data as an evidences to help its clients to make decision. This requires the mobile web service middleware should support WS-attachement in addition to basic SOAP messaging. Sharing multimedia data in a small group is highly required in epidemiological surveillance domain for clearer presentations and interactive discussions [13]. But multimedia content normally is much large than text content. It is unclear whether it is feasible to use handheld terminals to send/receive multimedia information as SOAP attachments in the mobile computing community.

In this study, we address the above challenges by developing a mobile web service (MWS) middleware. Particularly, we make the following contributions: (a) we design and develop a web services middleware for a collaborative mobile computing application so that each mobile terminal can be used as both a mobile service and a mobile service consumer; b) we propose a web services mobile framework (mBOSSS+) that allows an embedded device to cooperate and synchronize with peers for a class of mobile collaborative applications; (c) we demonstrate and evaluate the feasibility and performance of our solution by deploying and testing our MWS middleware using real mobile phones; (d) we also apply a performance model to predict the performance of mobile applications using our MWS middleware.

2   System Overview

In this section, we outline our MWS middleware and provide an in-depth description of the design of the MWS middleware.

2.1   MWS-Based Application Structure

An overview of the architecture is presented in Fig 1. Each handheld device publishes a WSDL description on a service broker. The web service description language (WSDL) describes the public interface to the web service. A mobile client in this network finds a request service(s) from the service broker. Then the mobile client retrieves the WSDL file to determine what operations are available on the mobile host. The mobile client uses SOAP via HTTP to call one of the operations listed in the WSDL. References [11] and [14] propose a way to make a mobile service broker, thus we assume there is a service broker that provides published WSDL to mobile requestors. Therefore, in this section, we focus our attention on the integration of the mobile host and the mobile requestors.

A detailed description of our middleware is presented in Fig 2. This architecture provides a model for developers to create flexible and reusable applications. It has been designed to support other types of applications that require large data sets to be stored and used on a handheld device, including those with wireless connectivity and with different displays and user interface capabilities. The architecture consists of four separate layers: an application processing tier, a data access tier, a data storage tier, and a mobile SOAP engine.

The application processing tier is consist of functional process logic and presentation of outputting results. Functional process logic will model business objects and describes the functional algorithms, then composes the logic to generate output.

The data access tier provides a data access interface (DAI), and a data centre (DC). The DAI provides a command interface for the database and isolates database operations (such as data retrieval) from data queries. This isolation allows the database structure to be adapted and modified independently of data query functionality. In other words, the DAI translates user requests into acceptable database commands. The DC is a mobile data cache with limited memory (in normal system the available memory was restricted to 5-10 MB). The DC provides the standard services required of a data cache and parts of database

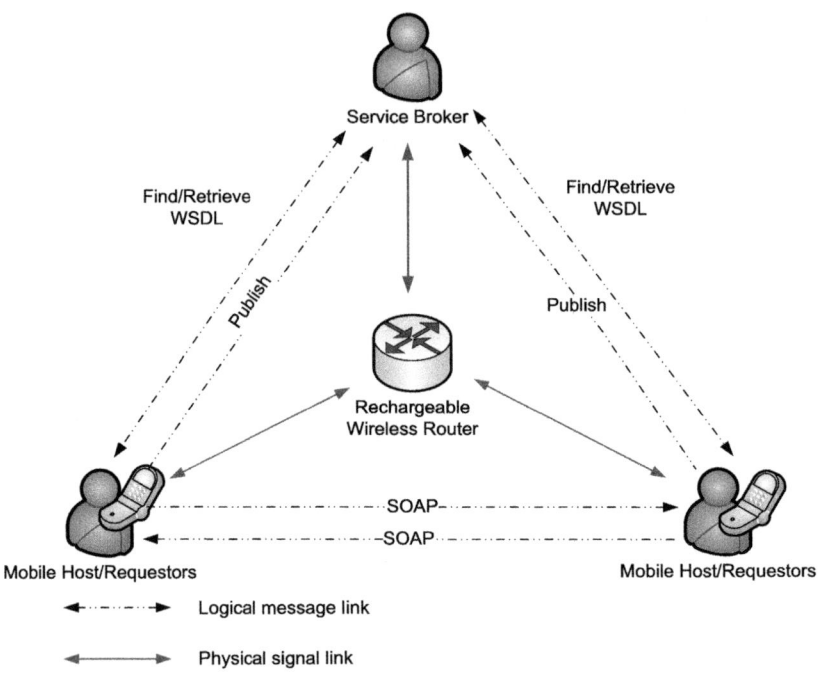

Fig. 1. Web services architecture in pervasive environment

interface functions such as database authentication, query parsing, and assembly of the search response.

An embedded database and a memory card database (MCD) comprise the MWS Storage system. The commercial embedded database can be used here for providing data query. The MCD stores the database tables as a system of folders and files on a SD memory card. The folders and files are arranged according to a holey-brick-N-ary (HB-Nary) search tree [15]. First, consider a single-attribute N-ary search tree. The data values of a single, specified database search key are coded as binary words. These binary words are then arranged according to a single-attribute N-ary search tree whose structure is based on a sequence of binary digits. The nodes of the N-ary search tree are then mapped to folders, where each folder corresponds to one particular search key value. The files inside each folder contain the table entries matching the search key value that has been mapped to that specific folder. Suppose now that we have two database search keys. In this case, we proceed as before for the first search key, only this time inside each folder is another set of folders that has been mapped to a single-attribute search tree for the second search key and so on. In this way, we construct a multi-attribute N-ary search tree for a particular database query. The holey-brick nature of the multi-attribute N-ary search tree simply indicates that pointers can be stored into data folders instead of the folders themselves. The pointers correspond to holes in the search tree and may be used to span multiple SD memory cards.

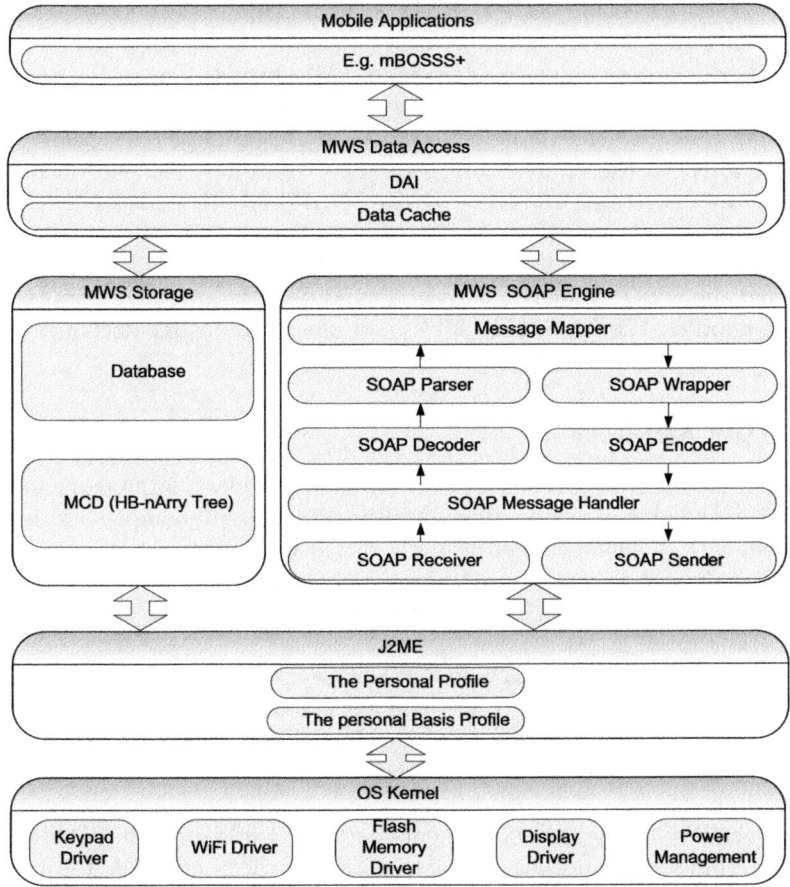

Fig. 2. The architecture of MWS middleware

The MCD stores data in a redundant format because there is one HB-Nary search tree for each specified data query. The essential feature of the MCD, however, is that it guarantees that the data files (table entries) inside each folder can be loaded entirely into the memory space of the DC. In other words, we have subdivided the tables of the database such that a portion of them can be loaded into memory and treated as an in-memory relational database by the DC.

The MWS SOAP engine consists of eight parts. The function of the message handler is to transfer request messages to the SOAP receiver and transfer SOAP messages from the SOAP sender. The Hypertext transfer protocol is used as a protocol to transfer messages. Messages come to the SOAP receiver, the SOAP decoder, and the SOAP parser and are then converted by the message mapper, which is responsible for serialization and deserializaition SOAP content to objects that a programming language can understand. The SOAP parser and the SOAP wrapper define the structure of a SOAP message with different elements and provide the functionality for parsing and wrapping the different elements of these messages. Then the SOAP decoder and SOAP encoder provide SOAP message encoding and the message decoding. These two parts define a way to encode/decode a data value to or from an XML format. One of the SOAP message handler function is to intercept messages from a mobile client to a mobile server and vice-versa. The handler supports encryption and decryption of SOAP messages for both the request and response. To request messages, the handler parses inbound messages and retrieves the SOAP body element for the SOAP decoder/encoder. To response messages, the handler constructs a complete SOAP message for the SOAP sender. The transport part is named in the SOAP receiver and the SOAP sender that receives and delivers messages from the SOAP message handler. HTTP, SMTP, FTP, and blocks extensible exchange protocol (BEEP) are accepted as a transport protocol.

2.2   MWS Attachment

The use of multimedia in a mobile environment provides a large scope for clearer presentations and discussions. Using multimedia content, complex stories can be described with a single image or a short video. Reference [16] describes the SOAP attachment as a feature. However, this SOAP attachment specification was not include in the SOAP messaging framework [17] and J2ME web services specification JSR172 [18].

Some techniques allow attachments to be sent on web services. The encapsulation of information in a SOAP attachment is a common use case and there are various ways in which this objective can be achieved, such as MTOM (SOAP message transmission optimization mechanism), SwA (SOAP with attachments), or base64binary.

The base64binary sends attachments as base64 inline in the SOAP message using pure binary. In another words, the attachment in embedded in the SOAP message. The disadvantages of pure binary are bloats the message by around 33% and converting the encapsulated data to a text format involves a high processing cost for decoding. The SwA sends the attachment outside of the SOAP message. A SOAP message can be composed into a multipurpose internet mail extensions (MIME) message. And the SOAP message contains a reference to the attachment. The MTOM describes a mechanism for optimizing the transmission or wire format of a SOAP message by selectively re-encoding portions of the message. A MTOM attachment is part of the SOAP message and allows the use of other attributes. MTOM messages can be signed and encrypted using WS-Security. Thus, this provides a mechanism to send secured attachments without the need for additional specs.

Reference [19] indicated that the SwA technique was the most efficient among the pure binary techniques, because it has the lowest response times in a high-bandwidth network. To achieve the best performance in mobile computing, we applied SwA as the protocol to send and receive SOAP attachments. Fig 3 presents the SOAP attachment feature properties used in the mobile framework. MWS request messages are transferred in an HTTP POST request and the response messages are transferred in an HTTP response message. In Fig 3, the

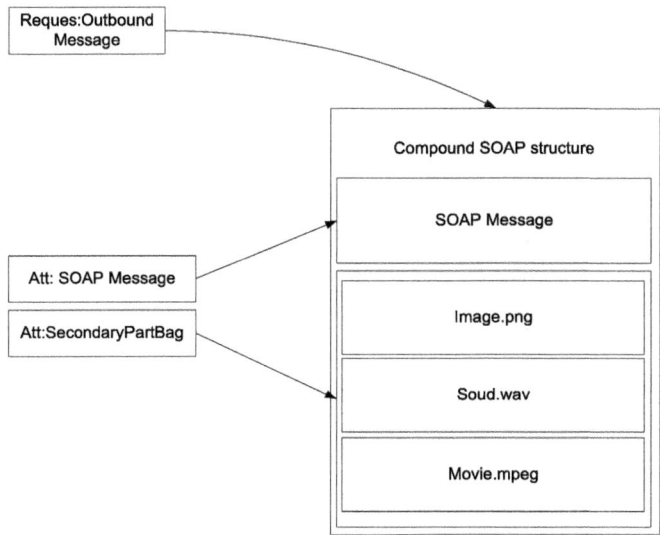

Fig. 3. Attachment feature properties

message type is text/xml which is used for the messages containing the SOAP envelope only. For MWS requests carrying a SOAP attachment, the content-type is multipart/related. The SOAP envelope is located in the first part of the MIME message along with the MWS attachments. MWS attachments are indicated by the start parameter of the multipart/related content-type. If an MWS attachment is included, it is encoded as a MIME part and is the second part of the HTTP post message. The MIME part should have the appropriate content type(s) to identify the payload. This MIME has two MIME headers; content-type and content-ID fields. The content-ID is referenced by the MWS request.

3   MWS Benchmark: Mobile Bovine Syndromic Surveillance System (mBOSSS)

Hand-held devices are generally not capable of handling very large amounts of data [6-7]; thus, they require mass data acquisition capacity and a system for providing the user with guidance that is adaptable. The mBOSSS+ aims to capture the observations of diseased cattle in a collaborative way.

Fig 4 presents a method to use the MWS middleware in a real application. The input/output tier aggregates the methods by which users interact with the application. There is a user interface, a communications interface with the operating system, and tools to connect with the network and other software. The user interface is an extremely important aspect of the mobile database application because there are no existing display-screen standards for mobile phones and other hand-held devices. While input methods are fairly standard (keypad, touch screen, and voice), the dimensions of display-screens vary greatly.

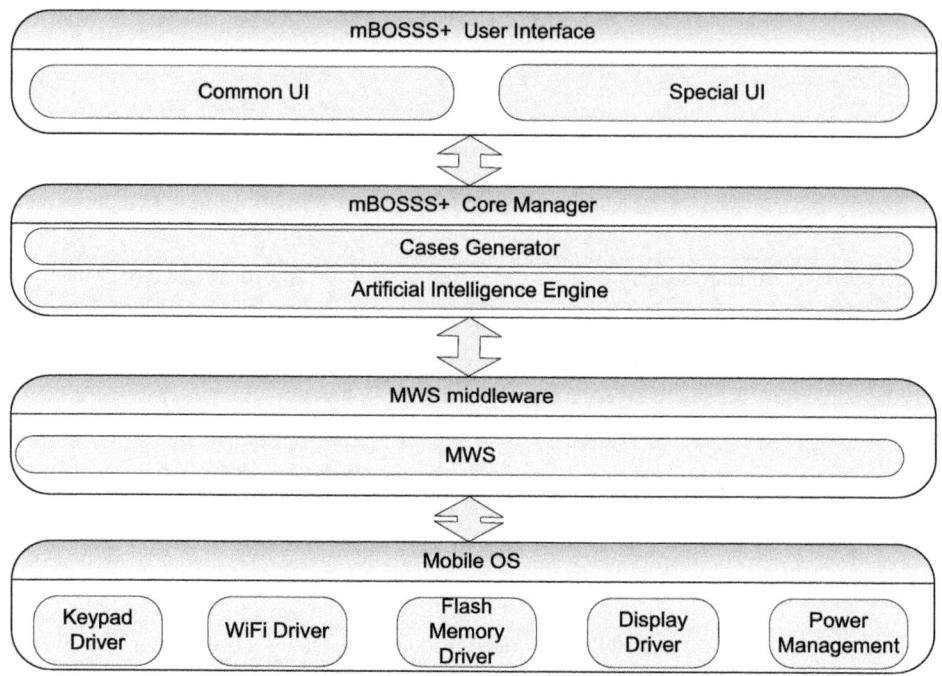

Fig. 4. WMS middleware + mBOSSS application

The core management tier encapsulates the core business model which provides the basis for all business transactions of the application. All syndromic surveillance and diagnostic cases are generated by the case generator. We applied a Naive Bayes classifier in an artificial intelligence engine. This tier determines the behavior of the application in response to different situations, including workflows and operating procedures. It generally includes the most complex components of the implementation.

The sequence diagram that corresponds to this model is presented in Fig 5 and depicts the interactions between cattle producers and the embedded system in sequential order of occurrence. There are two scenarios represented. The most common scenario is one in which the user investigates potential diseases that may result in the symptoms that they have observed in their cattle. The user enters observed symptom details into the device using the surveillance interface (drop-down menus and graphical aids). The system processes the data and provides the user with a list of possible diseases and offers suggestions to the user for further investigation. This includes asking the user questions about the presence or absence of specific symptoms. The information that is provided to the user is updated in an iterative process as new data is entered. This process can help to identify the disease with greater confidence and to improve the investigative pathway. The time savings provided by this field-enabled, real-time iterative process is the greatest advantage provided by the system.

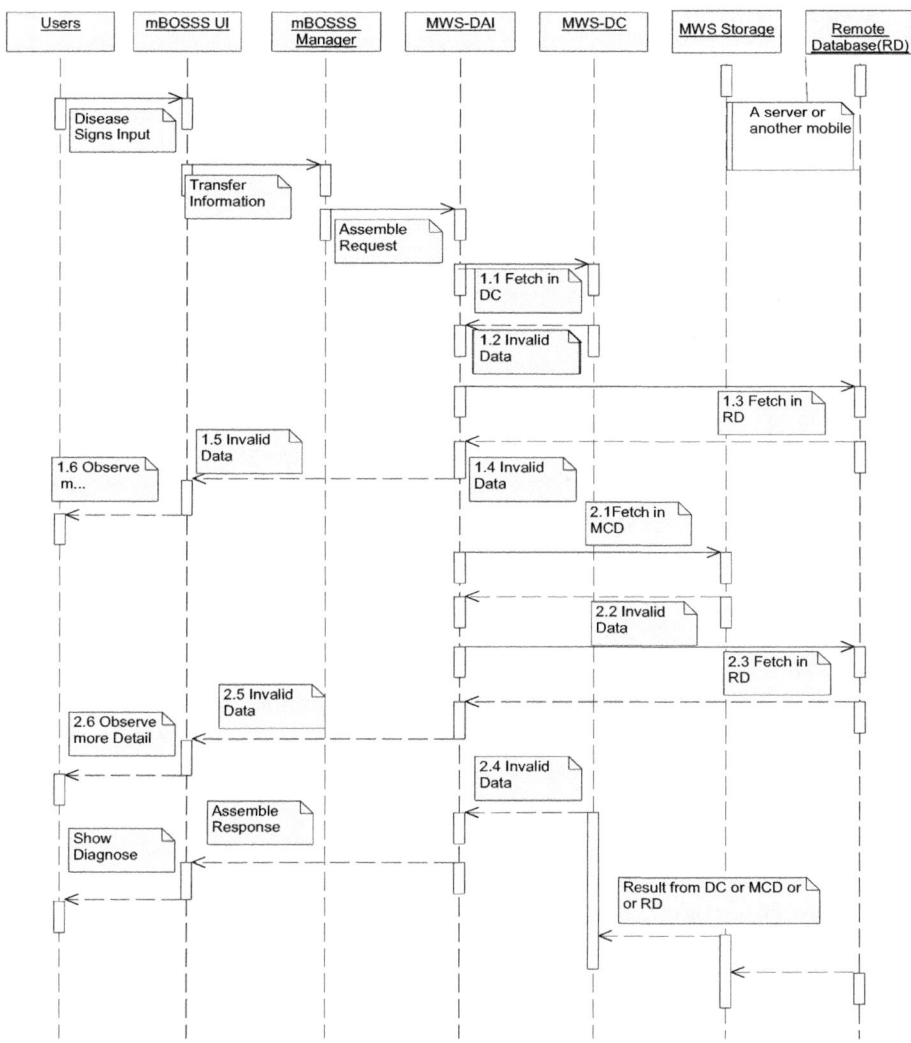

Fig. 5. The procedure of diagnostic a unhealthy cattle with mBOSSS+

In our diagnostic application, we employ a cattle disease database as sample data. This dataset of conditional probabilities contains approximately two millions probability estimates, a combination of nearly 3,000 diseases, and 2,000 signs. The table contains an average of 253 signs for each disease. The probability values are the result of complex calculations. Once the values are computed, the system can dynamically show the diagnoses and treat individual disease cases.

4   Performance Evaluation

We performed a series of experiments to assess the performance of our web services engine in this mobile application scenario. Performance is quite important to mobility software because the hardware cannot be easily updated, and the capability of handheld hardware is normally lower than that of a PC due to the size and weight requirements for the device. In our evaluation, the response time was measured as the performance metric between the time when the client mobile sent a request to the time when the client mobile received the corresponding response.

4.1   Case Selection

A mobile client selected four symptoms randomly from 3000 cases sent to a mobile device for disease diagnostic processes. After the diagnosis, three types of messages (Simple message, Medium message, Complex message) were designed to reflect common data structures used in our applications in terms of data size and complexity as follows:

- Simple message, four of the most likely diseases along with their descriptions without multimedia attachments for the symptoms provided.
- Medium message consists of four of the most likely diseases along with their descriptions and a 320x240 pixel picture to image the most likely diseases.
- Complex message, the four most likely diseases with their descriptions combined with four 320x240 pixel disease pictures.

4.2   Benchmark

We applied the cattle diseases database retrieved from the Australian Biosecurity Collaborative Research Centre as our benchmark dataset. We used a request/reply-style RPC model as our test scenario, which consisted of a client and a mobile web service. The client sends the web service a request with a list of animal disease symptoms. We design the request to contain information required for a test with the symptoms randomly selected from our dataset, which contains more than 3000 symptoms. The web service receives the request and replies by sending back a specific diagnosis with/without applying multimedia attachments according to the information in the request.

4.3   Test Environment and Settings

We conducted the performance tests in the following order. Both our web services middleware and the client are developed using Java platform, micro edition (J2ME). The web service is deployed and runs on a HTC 9500 mobile phone, which is running on IBM websphere everyplace micro environment that supports a connected device configuration (CDC1.1). The web service client runs on a J2ME (CDC1.1) simulator via a 108.11g battery-powered wireless router.

To ensure valid and accurate measurements, we ran each test multiple times and each test was run for 10 hours to guarantee consistent and stable test results. We also set 15% of the testing time at the beginning and end of each test as warm-up and cool-down periods, respectively, during which the testing results were not considered, to reduce the effect of unstable measurements during these periods. We also waited half a second between messages.

4.4 Testing and Observations

Fig 6 shows the average size of the Simple, Medium and Complex messages. The Simple messages included the text only, whose average message size was 1,280 Byte. However, once combined with multimedia data, the sizes dramatically increased 20-80 times for the Medium and the Complex messages. We compared the average sizes of the SOAP messages used in our performance tests in Fig 7.

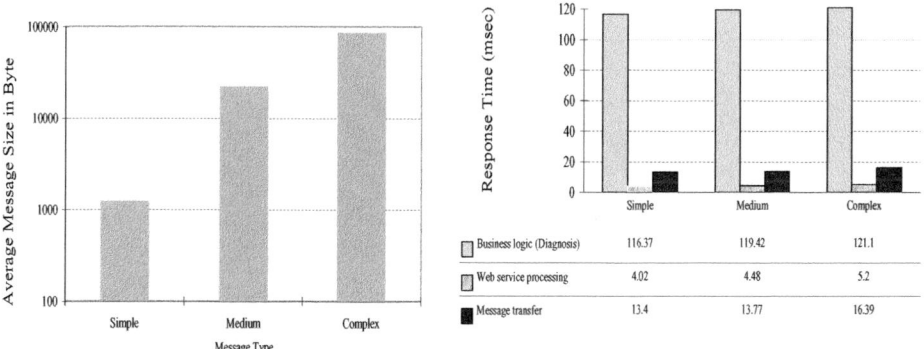

Fig. 6. Differences in size among message types

Fig. 7. Response time for the three message types

Fig 7 shows the time spent among the three activities (business logic, web service processing, and message transfer) on the mobile web service for the three types of messages. The business logic time included the local processing time spent searching for animal diseases and presenting the contents to the observers in the AIE (Artificial Intelligence Engine); the service provider processing time was the time spent preparing the SOAP responses according to the different message sizes and message types; and the service requester was the time spent processing SOAP requests, which included the time for messages transmission over a wireless network.

According to the test results shown in Fig 8, the shortest average response times was for simple SOAP messages, which increased, as expected, for more complex messages, and the multimedia based SOAP messages took the longest. The business logic activity costs most of the time, as it goes through the AIE and to a data retrieval process. The increasing time and message size trends are shown in Fig 8.

As the size of the message increased, the time difference spent among diagnostic time, service provider processing time, and service requester processing time increased. It can be concluded from this experiment that all times were affected by the size of the diagnostic message, but the increase in response time was much lower than the increase in message size. For example, the service provider size increased 68.7 times, but the service provider processing time increased by 6.6%. From our observations, 84%-87% of the time cost is spent on business logic. Thus, the total processing time does not increase as quickly as the message sizes grow.

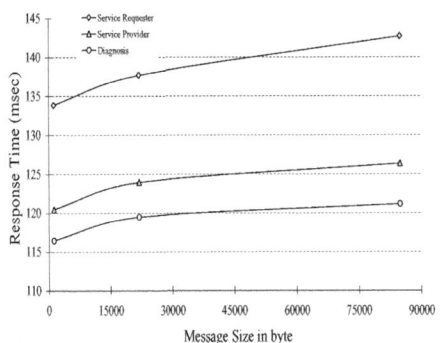 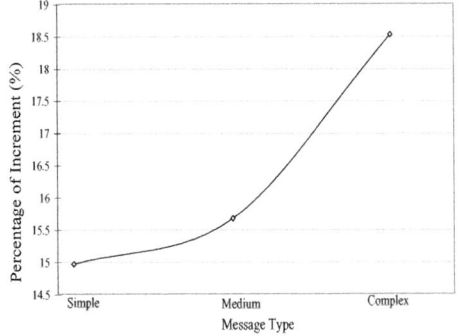

Fig. 8. Response time vs. message size in different stages

Fig. 9. Percentage of increment for the three message types

The percentage of response time increment between stand-alone mBOSSS and mBOSSS with web services features is shown in Fig 9. Hosting web services in mBOSSS results in slight time delays compared to the stand-alone version. The increases were between 15% and 19% and the values of delays were 17.42 ms, 18.24 ms, and 21.56 ms. This result indicates that using web services attachments for multimedia messages in mBOSSS is unlikely to have a large impact on system performance.

5   Modeling WMS Performance

In this section, the designed performance model is represented by the MWS middleware, by conduction the following performance analysis.

Fig 10 illustrates a method for making a remote call using SOAP (or TCP directly) in the wireless network reference model. First at the application level, a native data object needs to be serialised into a XML as a SOAP request. It takes considerable time to serialize data objects to XML. Then, the SOAP message is passed to the HTTP level. According to the standard HTTP protocol/configuration, the HTTP layer on the client-side needs to handshake with the service-side by sending a POST request. This request initiates a TCP connection between them. After receiving HTTP: 100 ACK, the client-side HTTP begins to send the entire SOAP message via TCP/IP. Note that while this handshake can be disabled, it is a default HTTP configuration and is relatively expensive even for small SOAP messages.

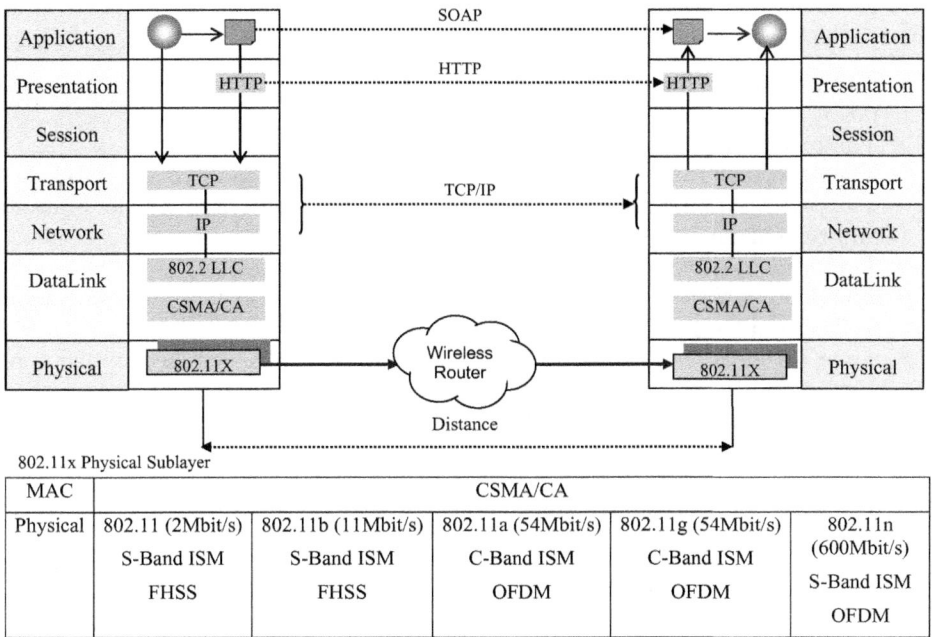

802.11x Physical Sublayer					
MAC	CSMA/CA				
Physical	802.11 (2Mbit/s) S-Band ISM FHSS	802.11b (11Mbit/s) S-Band ISM FHSS	802.11a (54Mbit/s) C-Band ISM OFDM	802.11g (54Mbit/s) C-Band ISM OFDM	802.11n (600Mbit/s) S-Band ISM OFDM

Fig. 10. Sending a SOAP request over a wireless network

The SOAP message may be partitioned into a set of small segments at the TCP layer. Appropriate headers and footers are attached to each segment as the segments are passed through transport, and network data link layers, until they reach the IEEE 802.11 physical sublayer. IEEE 802.11 defines a series of encoding and transmission schemes for wireless communications, such as Frequency Hopping Spread Spectrum (FHSS), Direct Sequence Spread Spectrum (DSSS), and Orthogonal Frequency-Division Multiplexing (OFDM) schemes. Then, there is a connection to a radio-based computer network via a wireless network interface controller (WNIC). The WNIC is responsible for putting the packages onto the wire at a specific speed (network bandwidth) relative to the next network device. During this period, overheads may result from network delays, including transmission delay, propagation delay, and processing delay on each WNIC.

The path from the bottom (physical layer) to the top (application layer) on the service-side is opposite to the process on the client-side: The received packages are unpacked at each layer and forwarded to the next layer for further retrieving. Note that as binary messages can be directly marshalled into native objects with less cost XML-based SOAP messages usually take more CPU cycles to be deserialised to native data objects. We assumed that the reverse path was symmetrical to the forward path, so the above analysis was not repeated for this case.

5.1   Performance Modeling

We extended our performance model [20-21] for our MWS middleware. In summary, the performance of the WSM depends on a number of factors, including:

- The processor time taken for workflow services. e.g., surveillance diagnosis process;
- The processor time for processing algorithms;
- The processor time for loading and writing a case from or to the database;
- The processor time taken to bind the SOAP messages carrying the requests and responses;
- The number of messages passed between the participants;
- The processor time taken to handle the transport protocols, such as HTTP, SMTP and TCP/IP;
- Time taken for each message to traverse the networks switches and routers;
- Time taken to travel the distance between the two mobiles at the speed of wireless devices.

Based on the above observations and analysis, we provide a simple static performance model for a typical request/reply call over the network as follows:

$$Y_{responsetime} = T_{msgProc} + T_{msgTran} + T_{synch} + T_{app} \tag{1}$$

Where:

- $T_{msgProc}$ represents the total cost of processing the messages, including coding or encoding, security checking, and data type marshalling;
- $T_{msgTran}$ represents the total cost to transfer a specific number of messages over the network;
- T_{synch} represents the overhead for the extra synchronization required by protocols;
- T_{app} represents the time spent in business logic at the application level.

For simplicity, we made the following assumptions:

- All network devices involved in the transmission have comparable capacity and, thus, no network overflow/retransmission occurred at any point.
- The message complexity was proportional to message size, and, thus, overhead for processing a message can be modeled into message sizes.

Based on the above assumptions, we modeled the three terms in (1) as follows:

Modelling $T_{msgProc}$. We modeled the overhead for processing messages as a linear function regarding message (data) size M_i:

$$T_{msgProc} = \sum_{i=1}^{\omega} \left[\sum_{j=1}^{2} (\alpha_j + \beta_j M_i) \frac{ref P_j}{P_j} \right] \tag{2}$$

Where:

- ω is the number of transits between the mobile client and mobile server, e.g., $\omega=1$ for one way sending, and $\omega=2$ for an normal request/response call;
- $refP_j$ specifies the CPU capacity of the reference platform;
- P_j represents the CPU capacity of the machine where client/server is deployed;
- α_j represents the identical inherent overhead for processing a client/server message built on specific middleware running on the reference platform;
- β_j represents the overhead for processing an unit number of messages (say 1KB) for the same middleware j also running on the reference platform.

Modelling $T_{msgTran}$. We modeled the overhead for transferring messages as follows:

$$T_{msgTran} = \sum_{i=1}^{\omega} \left[\sum_{j=1}^{n} (\tau_j + \frac{MoW_i}{N_j}) + \frac{D}{L} \right] \tag{3}$$

Where:

- n the total number of network devices involved;
- MoW the actual message size (byte) transferred on wire;
- N_j the bandwidth of the network devices (Mbps);
- τ message routing/switching delay at each ND;
- D the distance between the client and server;
- L the speed of a wireless communication is defined; as $200,000 km/s$. The frequency of wireless devices at which Wi-Fi standards 802.11b, 802.11g and 802.11n operate is 2.4 GHz. Due to electromagnetic interference that may interrupt, obstruct, or limit the effective performance of the circuit, the speed of light in water was used as the speed of wireless communication.

Modelling T_{synch}. Likewise, we modeled the overhead of the synchronizations that occurred during messaging as follows:

$$T_{synch} = \sum_{i=1}^{\varsigma} \left[\sum_{j=1}^{n} (\tau_j + \frac{m_j}{N_j}) + \frac{D}{L} \right] \tag{4}$$

Where:

- ς is the number of synchronizations occurring during messaging and defined as:

$$\varsigma = \delta(http) + \left\lceil \frac{MoW}{W_s} \right\rceil \tag{5}$$

- m the message size for each synchronization;
- W the TPC window size ranging from 16K to 64K.

Modelling T_{msgApp}. We modeled the time spent on business logic as a function regarding to message (data) size M_i:

$$T_{msgApp} = \sum_{i=0}^{\alpha} T_{processing} + \sum_{i=0}^{\rho} T_{operation} \tag{6}$$

Where:

- α is the number of process algorithms serviced for one diagnostic process;
- ρ is the number of operations being invoked;
- $T_{processing}$ is the time spent on processing algorithms;
- $T_{operation}$ is the time cost on data operation including data search, data load, data written into storage, or removing data from storage.

$$T_{operation} = \begin{cases} T_{search} \\ T_{load} \\ T_{write} \\ T_{remove} \end{cases} \tag{7}$$

- T_{search} is the time spent identifying the data;
- T_{load} is the time spent loading data; from data storage;
- T_{write} is the time spent on data updates;
- T_{remove} is the time spent removing an item (e.g., symptom or disease) from storage.

Put it all together. With the above modeling and making a call over the network, we can provide a generic performance model for distributed calling and/or messaging:

$$
\begin{aligned}
Latency = & \sum_{i=1}^{\alpha} T_{processing} + \sum_{i=1}^{\rho} T_{operation} + \sum_{i=1}^{\varsigma}\left[\sum_{j=1}^{n}(\tau_j + \frac{m_j}{N_j}) + \frac{D}{L}\right] \\
& + \sum_{i=1}^{\omega}\left[\sum_{j=1}^{n}(\tau_j + \frac{MoW_i}{N_j}) + \frac{D}{L}\right] + \sum_{i=1}^{\omega}\left[\sum_{j=1}^{2}(\alpha_j + \beta_j M_i)\frac{refP_j}{P_j}\right]
\end{aligned} \tag{8}
$$

While this model is complex and considers heterogeneities in a general case, it can be significantly simplified in a real environment and usages. Let us discuss how to simplify (8) by make some reasonable assumptions:

- Set $m_i/N_j = 0$
  As the size of the HTTP synchronization message, i.e., m_i in (4) is very small, this means that m_i/N_j tends to be zero in most cases and, thus, can be omitted.
- Treat $\tau_j = 0$ as a constant
  Although $\tau_j = 0$ is a variable for different network devices and related to the processor capacity inside the network devices (very difficult to get in practice), its value is very small and should not change greatly based on current hardware. Therefore, we can treat it as a constant.

– Make N_j as a constant
All network devices (switches/routers) have the same bandwidth, so N_j is a constant.
– Set $T_{processing}$ and $T_{operation}$ as a constant
The time costs on software systems are different from one to another, so the values are considered constants. In following mode we set time to zero.

As a result, our performance model (8) becomes:

$$Latency = \sum_{i=1}^{\varsigma}\left[n\tau + \frac{D}{L}\right] + \sum_{i=1}^{\omega}\left[\sum_{j=1}^{n}(\tau_j + \frac{MoW_i}{N_j}) + \frac{D}{L}\right]$$
$$+ \sum_{i=1}^{\omega}\left[\sum_{j=1}^{2}(\alpha_j + \beta_j M_i)\frac{refP_j}{P_j}\right] \tag{9}$$

The performance mode (9) can be further simplified based on our specific test scenario and platforms:

– All calls are two ways, i.e. $\omega = 2$;
– The client and server applications are built on the same middleware, i.e. $\alpha 1 = \alpha 2 = \alpha$ and $\beta 1 = \beta 2 = \beta$
– The performance parameters (α, β) are obtained on the same reference platform, i.e. $refP1 = refP2 = refP$.

Therefore, the performance model used in this report is as follows:

$$Latency = \sum_{i=1}^{\varsigma}\left[n\tau + \frac{D}{L}\right] + \sum_{i=1}^{\omega}\left[(n\tau + n\frac{MoW_i}{N}) + \frac{D}{L}\right] + \sum_{i=1}^{\omega}2(\alpha + \beta M_i) \tag{10}$$

Parameter fitting. This section demonstrates how to fit the parameters in the models that we developed in the last section. To validate our performance model, we intended to use different tests from our previous tests to fit the parameters as closely as possible.

Fitting τ. To isolate τ from other parameters in (10), we applied our middleware to measure the latency of a sending/receiving round-trip for very small messages, say 1 byte. In this test, we made the following reasonable approximations for simplicity:

The overhead for processing data/message can be ignored, i.e., $T_{processing} = 0$, due to small message size and using socket (with little overhead at middleware level).

The overhead for transferring data/message is very small, i.e., $MoW/N = 0$, due to a small message (say 1 byte) over a very fast network ($100Mbps$). As a result, the transmission overhead (3) becomes:

$$T_{msgTran} = 2(n\tau + \frac{D}{L}) \tag{11}$$

We measured the latency of 100 round trips by running the socket program over LAN and obtained an average latency: $T_{t}ransmission = 0.36msec$. Because we know that there are three network devices (including client and server machines) along the network path, i.e. $n = 3$, we obtain τ from (11) as follows:

$\tau = (T_{msgTran}/2D/L)/n = (0.36/20.01/200)/3 \approx 0.06msec$

Note that the network parameter τ is a variable for different network devices at runtime, depending on many static factors of the devices/machines and dynamic states of the network. In the remainder of this paper, we will use the same $\tau = 0.06$ for all performance predictions.

Fitting $\alpha and \beta$. Because the two parameters are related to message processing rather than networking, they can be derived from (2) using the standard least-squares fitting method. We applied the Utility Law [22] to obtain the time spent processing messages as follows:

$$T_{msgProc} = \frac{\lambda}{Throughout} \tag{12}$$

As a result, fitting α and β from (2) becomes a typical linear regression problem. We will use MWS middleware as an example to demonstrate how it works. First, we ran a set of tests using Medium message definition with 1, 20, 40, 60 symptoms, respectively. We calculated the processing times and message sizes for each test run with (2) as follows:

Taking the data set $(M$ and $Tp/2)$ in the above table as input, we derived $\alpha and \beta$ with S-Plus [12] as follows:

- $\alpha = 1.05$
- $\beta = 0.36$

5.2   Model Validation

App vs. nonApp. First, we compared our middleware with an application and the system without an application. The test results with the latencies derived from our performance model (10) were compared with the parameters obtained from the last subsection. The comparison results are summarized in Fig 11 (a) and (b), respectively.

As shown in Fig 11, the latencies predicted with our performance model fit the latencies that we measured very well for the MWS middleware across the three messages (Simple, Medium and Complex). The prediction errors are all under 12%. Therefore, it is feasible to use our performance model to prediction SOAP performance for a large range of messages types and sizes. It also demonstrates that our performance model is not only feasible for SOAP, but is also a general-purpose distributed systems performance model that is applicable to common request/response messaging built on different technologies.

IEEE 802.11a vs. IEEE 802.11b: We directly applied the derived parameters (α and β) to predict the performance of both IEEE 802.11a and IEEE 802.11g. The predicted and experimental results are shown in Fig 11 (c) and Fig 11 (d).

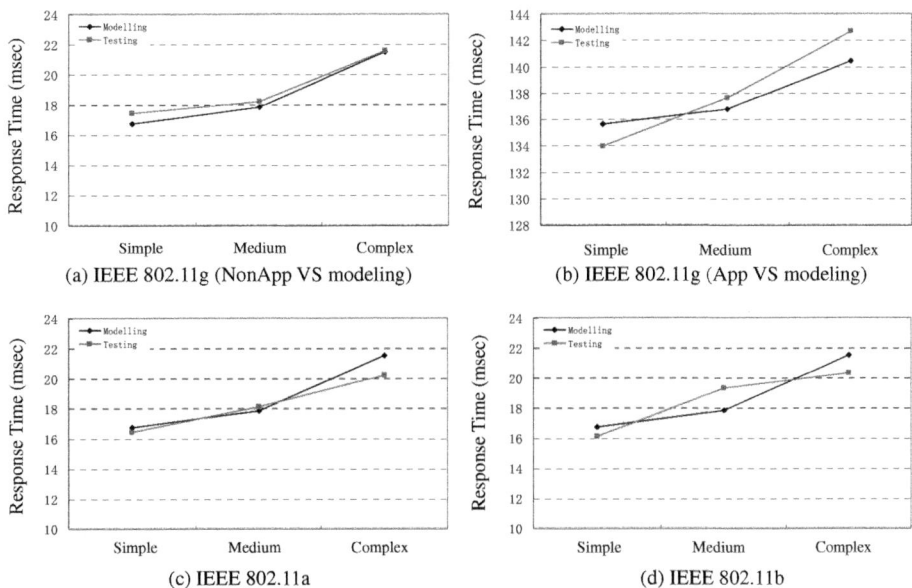

Fig. 11. Model validation with test scenarios

This figure indicates our performance isolates for different overheads well. So far, from all the above results, we have shown that our performance model, with the parameters, successfully predicted SOAP performance over Wireless LAN by reusing most model parameters. The validation results demonstrate the ability of our performance model to handle Wi-Fi networks.

6   Related Work

We discuss the most closely related work in this section. Several studies on enabling web services on a mobile system have been conducted.

Asif et al. [9-10] proposed a lightweight web services engine, which includes an protocol listener and a security subsystem for providing web services on handheld devices. Srirama et al. [11-12,25] detailed a mobile web service mediation framework (MWSMF) for efficiently providing web services in a mobile P2P network. They used a binary compression technology (BinXML) for efficient communication. Chatti et al. [23] presented a smart phone driven mobile web services architecture for collaborative learning. This paper emphasized the importance of collaboration, and knowledge for learning. [23] described the integration of mobile and ubiquitous technologies to enhance collaborative learning activities and present the details of smart phone sent text-based SOAP messages as an approach for mobile learning. Kim and Lee [24] and Pham and Gehlen [14] proposed two web service platforms for mobile devices. The two frameworks contain several built-in functionalities such as processing of SOAP messages, execution and migration of services and context management. These frameworks implement lightweight SOAP server architectures for Java2 micro edition (J2ME) devices. The implementations of web service parts for mobile devices are based on the opensource package kSOAP [28] for client function and they used a opensource software kXML[29] for the xml parser. As described above, those papers only investigated the feasibility of hosting web services functions on handheld devices. But the architecture and the algorithm for delivering web services with attachments to the mobile computing domain have not been addressed.

Srirama et al. [12] studied the performance of their mobile SOAP implementations. They created a prototype with J2ME and KSOAP to test time delays for a web service provider. Chatti et al. [23] evaluated their framework with text-based learning materials. Those studies did not include any performance modeling contents about mobile web services. The middleware performance model in this paper provides guidance for performance cost as deploying web service on mobile devices. Moreover, there are few studies that have evaluated the scalability and performance of attached multimedia information with SOAP messages for mobile web services.

Mobile social network is becoming another popular mobile application. In mobile network, people exchange their location date that allows users to locate one another. Zheng et al. [30-31], [36] present a social-networking service called GeoLife. GeoLife aims to put the information generated by GPS-enabled mobile devices to share life experiences and build connections among each other using human location history. [32] proposes a mobile based social networking service that describes a tuple space model located in mobile devices. Our mobile web service framework proposed in this paper facilitates this model by enabling its self defined connection protocols with a mobile web service.

On the server side, some studies using Apache Axis [33] as a SOAP server with Apache Tomcat [34] as a application server or Microsoft SOAP have been evaluated and overall performance and scalability [19], [26-27], [35]. These studies presented performance models to predict the SOAP response time. Their models considers some factors, including compression methods, maximum segment size, round trip time, initial time-out sequence value, the number of packets per ACK, maximum congestion control window size and packet loss rate, which are neither necessary nor practical in real applications. In addition, their model validation is based third parties middleware on simulated data instead of experimental results, which are different from our studies. We obtained from real mobile devices as demonstrated in this study. Therefore, the result and model presented in this paper are more valid and convince.

In these related approaches, few of them can enable delivering web services on a mobile system. Meanwhile, while middlewares [9-12], [23-25] support mobile web services, their mobile web services do not provided web services attachments. [19], [26-27], [35] present their performance evaluation and prediction using the SOAP response time. However, these evaluations are conducted on the server side, which does not reflect end-to-end overhead in a wireless network. Our mobile web service middleware supports both standard SOAP protocol, as well as web service attachment specification, which enhances the compatibility of our mobile middleware so that it can be used in a wider range of mobile applications. Our mobile web service framework has been prototyped and deployed onto a real mobile phone, where our performance experiments were conducted. Therefore, our experiment results reflect the actuarial overheads between a mobile web service and its mobile clients.

7   Conclusion

In this paper, we have presented a complete research and development cycle of MWS middleware motivated by a real mobile epidemiological surveillance and diagnosis application. We provided the design, implementation, evaluation, and performance prediction of our mobile web service middleware, which facilitates mobile web services in a distributed mobile environment. This work contributes to web services on mobile distributed computing and allows mobile devices to exchange information as multimedia SOAP messages within SwA (SOAP with Attachment). Our WMS middleware consists of four core components: a soap engine, a storage model, a data access tier, and an application processing tier.

Besides a detailed description of our middleware, we applied our MWS middleware for a mobile application (mBOSSS+) that utilized this middleware to surveil and diagnose cattle diseases in the field and is designed to deliver benefits to cattle producers, government surveillance systems and the cattle industry. An experimental evaluation of our real-world workload shown that our middleware and application have acceptable overhead for transferring multimedia materials in such a wireless environment.

Based on our evaluations and observations, we presented a performance model for mobile web service applications. This mobile performance model can be used by mobile web services architects and mobile application developers who need to predict performance during the analysis and design phases.

For the future work, we plan to apply our MWS to a broad range of applications, such as location-based social networking services [30-31]. A handheld device with MWS will share its movement trajectories to build connections among each other using location history. In addition, applications of our middleware in other domains will also be investigated.

Acknowledgements

This work is funded by Australian Biosecurity Collaborative Research Centre (Grant No. 3.015Re).

References

1. Berners-Lee, T.: Twenty Years: Looking Forward, Looking Back. In: International World Wide Web Conference (2009),
http://www.w3.org/2009/Talks/0422-www2009-tbl/#(3)
(Accessed May 6, 2010)

2. Balan, R.K., Ramasubbu, N., Prakobphol, K., Christin, N., Hong, J.: mFerio: the design and evaluation of a peer-to-peer mobile payment system. In: The 7th International Conference on Mobile Systems, Applications, and Services, pp. 291–304 (2009)
3. Zhang, J., Levy, D., Chen, S.: A Mobile Learning System For Syndromic Surveillance And Diagnosis. In: The 10th IEEE International Conference on Advanced Learning Technologies, pp. 92–96 (2010)
4. Giurgiu, I., Riva, O., Juric, D., Krivulev, I., Alonso, G.: Calling the cloud: Enabling mobile phones as interfaces to cloud applications. In: Bacon, J.M., Cooper, B.F. (eds.) Middleware 2009. LNCS, vol. 5896, pp. 83–102. Springer, Heidelberg (2009)
5. Garnerin, P., Valleron, A.: The French computer network for the surveillance of communicable diseases. In: The Annual International Conference of the IEEE, pp. 102–107 (1988)
6. Kephart, J.O., White, S.R., Chess, D.M.: Computers and epidemiology. IEEE Spectrum 30(5), 20–26 (1993)
7. Vourch, G., Bridges, V., Gibbens, J., Groot, B., McIntyre, L., Poland, R.: Detecting Emerging Diseases in Farm Animals through Clinical Observations. Emerg Infect Diseases 12(4), 450–460 (2006)
8. Zhang, J., Calvo, R., Jin, C., Sherld, R.: A Framework for Mobile Disease Report and Investigation. In: International Conference on Mobile Technology Applications and Systems, pp. 59–62 (2006)
9. Asif, M., Majumdar, S., Dragnea, R.: Partitioning the WS Execution Environment for Hosting Mobile Web Services. In: IEEE International Conference on Services Computing, vol. 2, pp. 315–322 (2008)
10. Asif, M., Majumdar, S., Dragnea, R.: Hosting Web Services on Resource Constrained Devices. In: IEEE International Conference on Web Services, pp. 583–590 (2007)
11. Srirama, S., Vainikko, E., Sor, V., Jarke, M.: MWSMF: a mediation framework realizing scalable mobile web service provisioning. In: The 1st International Conference on Mobile Wireless Middleware, pp. 1–7 (2007)
12. Srirama, S.N., Jarke, M., Zhu, H., Prinz, W.: Scalable Mobile Web Service Discovery in Peer to Peer Networks. In: The Third International Conference on Internet and Web Applications and Services, pp. 668–674 (2008)
13. Pobiner, S.: Collaborative multimedia learning environments. In: CHI 2006 Extended Abstracts on Human Factors in Computing Systems, pp. 1235–1240 (2006)
14. Pham, L., Gehlen, G.: Realization and Performance Analysis of a SOAP Server for Mobile Devices. In: The 11th European Wireless Conference, vol. 2, pp. 791–797 (2005)
15. Zhang, J.: A Mobile Surveillance Framework for Animal Epidemiology. The University of Sydney (2008)
16. W3C: SOAP 1.2 Attachment Feature. In: W3C (2004), http://www.w3.org/TR/soap12-af/
17. W3C: SOAP Version 1.2. In: W3C (2007), http://www.w3.org/TR/soap12-part1/
18. Java Community: JSR 172: J2METM Web Services Specification. In: Java Community Process (2004), http://jcp.org/en/jsr/detail?id=172
19. Estrella, J.C., Endo, A.T., Toyohara, R.K.T., Santana, R.H.C., Santana, M.J., Bruschi, S.M.: A Performance Evaluation Study for Web Services Attachments. In: IEEE International Web Services, pp. 799–806 (2009)
20. Chen, S., Yan, B., Zic, J., Liu, R., Ng, A.: Evaluation and Modeling of Web Services Performance. In: IEEE International Conference on Web Services, pp. 437–444 (2006)

21. Chen, S., Liu, Y., Gorton, I., Liu, A.: Performance prediction of component-based applications. J. Syst. Softw. 74(1), 35–43 (2005)
22. Menasce, D.A., Almeida, V.A.F.: Scaling for E-Business: Technologies, Models, Performance, and Capacity Planning. Prentice-Hall, Englewood Cliffs (2000)
23. Chatti, M.A., Srirama, S., Kensche, D., Cao, Y.: Mobile Web Services for Collaborative Learning. In: The Fourth IEEE International Workshop on Wireless, Mobile and Ubiquitous Technology in Education, pp. 129–133 (2006)
24. Kim, Y.-S., Lee, K.-H.: A lightweight framework for mobile web services. Computer Science - Research and Development 24(4), 1865–2042 (2009)
25. Srirama, S., Vainikko, E., Sor, V., Jarke, M.: Scalable Mobile Web Services Mediation Framework. In: The Fifth International Conference on Internet and Web Applications and Services, pp. 315–320 (2010)
26. Davis, D., Parashar, M.: Latency Performance of SOAP Implementations. In: IEEE Cluster Computing and the GRID, pp. 407–410 (2002)
27. Punitha, S., Babu, C.: Performance Prediction Model for Service Oriented Applications. In: 10th IEEE International Conference on High Performance Computing and Communications, pp. 995–1000 (2008)
28. KSOAP2 project, http://ksoap2.sourceforge.net/ (accessed June 6, 2010)
29. KXML2 project, http://kxml.sourceforge.net/ (accessed June 6, 2010)
30. Zheng, Y., Chen, Y., Xie, X., Ma, W.-Y.: GeoLife2.0: A Location-Based Social Networking Service. In: Tenth International Conference on Mobile Data Management: Systems, Services and Middleware, MDM 2009, May 18-20, pp. 357–358 (2009)
31. Zheng, Y., Xie, X., Ma, W.-Y.: GeoLife: A Collaborative Social Networking Service among User, location and trajectory. IEEE Data(base) Engineering Bulletin 2(33), 32–39 (2010)
32. Sarigol, E., Riva, O., Alonso, G.: A tuple space for social networking on mobile phones. In: 2010 IEEE 26th International Conference on Data Engineering (ICDE), March 1-6, pp. 988–991 (2010)
33. Apache project, http://ws.apache.org/axis/ (accessed June 6, 2010)
34. Tomcat project, http://tomcat.apache.org/ (accessed June 6, 2010)
35. Liu, H., Lin, X., Li, M.: Modeling Response Time of SOAP over HTTP. In: IEEE International Conference on Web Services, pp. 673–679 (2005)
36. Zheng, Y., Xie, X.: Learning travel recommendations from user-generated GPS traces. ACM Transaction on Intelligent Systems and Technology (ACM TIST) 2(1), 2–19 (2011)

Integrating Large and Distributed Life Sciences Resources for Systems Biology Research: Progress and New Challenges

Hasan Jamil

Department of Computer Science
Wayne State University, USA
jamil@cs.wayne.edu

Abstract. Researchers in Systems Biology routinely access vast collection of hidden web research resources freely available on the internet. These collections include online data repositories, online and downloadable data analysis tools, publications, text mining systems, visualization artifacts, etc. Almost always, these resources have complex data formats that are heterogeneous in representation, data type, interpretation and even identity. They are often forced to develop analysis pipelines and data management applications that involve extensive and prohibitive manual interactions. Such approaches act as a barrier for optimal use of these resources and thus impede the progress of research.

In this paper, we discuss our experience of building a new middleware approach to data and application integration for Systems Biology that leverages recent developments in schema matching, wrapper generation, workflow management, and query language design. In this approach, ad hoc integration of arbitrary resources and computational pipeline construction using a declarative language is advocated. We highlight the features and advantages of this new data management system, called *LifeDB*, and its query language *BioFlow*. Based on our experience, we highlight the new challenges it raises, and potential solutions to meet these new research issues toward a viable platform for large scale autonomous data integration. We believe the research issues we raise have general interest in the autonomous data integration community and will be applicable equally to research unrelated to LifeDB.

1 Introduction

In systems biology research, data and application integration is not a choice – it is critical and has been identified as the major obstacle toward harvesting the true potential of the post genomic era life sciences research. It is complicated by the multiple different axes of integration considerations such as (i) schema heterogeneity (both syntactic and semantic), (ii) data heterogeneity (traditional flat data versus long strings of DNA or protein, structure data such as protein primary, secondary and tertiary structures, network data such as protein-protein, protein-gene, protein-RNA, gene-gene interaction, pathway data such

A. Hameurlain, J. Küng, and R. Wagner (Eds.): TLDKS III , LNCS 6790, pp. 208–237, 2011.

as metabolic and signalling pathways, expression data such as gene, protein and tissue expression, and so on, and flat versus nested data as in XML), and (ii) physical location of resource repositories (either locally or as deep web resources). Considerable effort has been devoted in the past to address schema heterogeneity [69,63,61] in traditional data management research. In semantic data integration domain and artificial intelligence, record linkage [56,71,64] and information aggregation [65,37] have seen progress in recent years. In more core data integration research, ontology generation or schema matching [39,20,68], and wrapper generation [58,29,19] have been explored extensively. Unfortunately, until recently, these interrelated components have not been assembled in any single data management platform to offer a comprehensive solution to this pressing problem.

It is our observation that most high profile life sciences resources, with a few notable exceptions, are need driven implementations that were developed mostly by a biologist with simplicity of use in mind which often led to the form based graphical interfaces, key word based search and hyper link based hierarchical information access. These resources were also developed in the pre-genomic era when large scale systems biology research was common. In the post genomic era when massive data sets are being produced by high throughput systems such as mass spec, gene expression, whole genome sequencing and new generation sequencing, the traditional interface driven and link based access to data has become a limiting factor. Yet, the popularity and simplicity of these systems continue to guide modern implementations, so much so that many high profile public repositories such as GenBank, PDB, GO, etc. continue to support this approach in addition to offering data download in large volumes.

2   Characteristics and Uniqueness of Data Integration in Life Sciences

In the absence of a single data management system, researchers took advantage of the publicly as well as privately available resources (that they could download and maintain if needed) and approached data integration from in multiple different directions. The current state of the sharable online life sciences resources have been in the making for about two decades. These resources initially were designed to serve specific application needs without any concerted thought to sharing on a larger scale, technology choice, or its impact on the maintenance of the systems. In these traditional approaches, data and tools for interpreting them from multiple sources are warehoused in local machines, and applications are designed around these resources by manually resolving any existing schema heterogeneity. This approach is reliable, and works well when the application's resource need, or the data sources do not change often, requiring partial or full overhauling. The disadvantage is that the warehouse must be synchronized constantly with the sources to stay current, leading to huge maintenance overhead. The alternative has been to write applications using dedicated communication with the data sources, again manually mediating the schema. While this approach removes physical downloading of the source contents and buys currency, it still requires manual mediation, coping with changes in the source, and writing source specific glue codes that cannot be reused. The basic assumption here is that the sources are autonomous and offer a "use as you see" and hands off support. That means that the application writer receives no support in any form or manner from sources other than the access.

There has been a significant effort to alleviate the burden on application writers for this alternative approach by developing libraries in popular scripting languages such as Perl and PHP for accessing and using resources such as GenBank, UCSC, PDB etc., and facilitating application development using the sources for which tested scripts are available. Consequently, applications that demand change, access to new resources, are transient or ad hoc, and are not ready to commit to significant maintenance overhead remain ill-served. The optimism, however, was short lived (though people still use this approach) because more interesting and more informative resources keep emerging and the dependence on standard resources such as GenBank and UCSC quickly became limiting as researchers struggled to adopt the newer resources for their applications. This approach also did not address the heterogeneity of the databases and the data types in the databases, and thus users were left to resolve those heterogeneities manually, forcing them to be knowledgeable at the systems level. As discussed earlier, biologists never really embraced this way of interacting [35], where they would have to learn programming as a tool.

In this paper, we introduce a new approach to on-the-fly autonomous information integration by way of examples that removes several of the hurdles in accessing life sciences resources at a throw away cost, and without any need for strict coupling or dependence among the sources and the applications. The new data management system we introduce is called *LifeDB* [23] offering a third alternative that combines the advantages of the previous two approaches – currency and reconciliation of schema heterogeneity, in one single platform through a declarative query language called *BioFlow* [54]. In our approach, schema heterogeneity is resolved at run time by treating hidden web resources as a virtual warehouse, and by supporting a set of primitives for data integration on-the-fly, to extract information and pipe to other resources, and to manipulate data in a way similar to traditional database systems. At the core of this system are the schema matching system *OntoMatch* [24], the wrapper generation system *Fast-Wrap* [19], and a visual editor called *VizBuilder* [48], using which users are able to design applications using graphical icons without the need for ever learning BioFlow for application design. In BioFlow, we offer several language constructs to support mixed-mode queries involving XML and relational data, workflow materialization as processes and design using ordered process graphs, and structured programming using process definition and reuse. We introduce LifeDB's architecture, components and features in subsequent sections. However, we will not discuss FastWrap, OntoMatch and VizBuilder in any detail although they are integral components of LifeDB. We refer interested readers to their respective published articles in the literature.

3   LifeDB Data Integration System: BioFlow by Example

To illustrate the capabilities of LifeDB, we adapt a real life sciences application discussed in [41] which has been used as a use case for many other systems and as such can be considered a benchmark application for data integration. A substantial amount of glue codes were written to implement the application in [41] by manually reconciling the source schema to filter and extract information of interest. Our goal in this section is to show how simple and efficient it is to develop this application in LifeDB.

The query, or workflow, the user wants to submit is the hypothesis: *"the human p63 transcription factor indirectly regulates certain target mRNAs via direct regulation of miRNAs"*. If positive, the user also wants to know the list of miRNAs that indirectly *regulate* other target mRNAs with high enough confidence score (i.e., $pValue \leq 0.006$ and $targetSites \geq 2$), and so he proceeds as follows. He collects 52 genes along with their chromosomal locations (shown partially in figure 1(a) as the table *genes*) from a wet lab experiment using the host miRNA genes and maps at or near genomic p63 binding sites in the human cervical carcinoma cell line ME180. He also has a set of several thousand direct and indirect protein-coding genes (shown partially in figure 1(d) as the table *proteinCodingGenes*) which are the targets of p63 in ME180 as candidates. The rest of the exploration proceeds as follows.

miRNA	chromosome
hsa-mir-10a	ch 17
hsa-mir-205	ch 1

(a) genes

microRNA	geneName	pValue
hsa-mir-10a	FLJ36874	0.004
hsa-miR-196b	MYO16	0.009

(b) sangerRegulation

geneID	miRNA	targetSites	pValue
FLJ36874	hsa-mir-10a	10	0.004
FLJ36874	hsa-mir-10b	3	*null*
RUNDC2C	hsa-mir-205	8	*null*
MYO16	hsa-miR-196b	*null*	0.009

(e) regulation

geneID	miRNA	targetSites
FLJ36874	hsa-mir-10a	10
FLJ36874	hsa-mir-10b	3
RUNDC2C	hsa-mir-205	8

(c) micrornaRegulation

Gene	p63Binding
FLJ36874	Y
RUNDC2C	Y
MYO16	N

(d) proteinCodingGene

geneID	miRNA	targetSites	pValue	p63Binding
FLJ36874	hsa-mir-10a	10	0.004	Y
FLJ36874	hsa-mir-10b	3	*null*	Y
RUNDC2C	hsa-mir-205	8	*null*	Y
MYO16	hsa-miR-196b	*null*	0.009	N

(f) proteinCodingGeneRegulation

Fig. 1. User tables and data collected from microRNA.org and microrna.sanger.ac.uk

He first collects a set of genes (*geneIDs*) for each of the miRNAs in the table *genes*, from the web site www.microrna.org by submitting one miRNA at a time in the window shown in figure 2(a), that returns for each such gene, a set of gene names that are known to be targets for that miRNA. The site returns the response as shown in figure 2(b), from which the user collects the *targetSites* along with the gene name partially shown as the table *micrornaRegulation* in figure 1(c).

To be certain, he also collects the set of gene names for each miRNA in table *genes* from microrna.sanger.ac.uk in a similar fashion partially shown in table *sangerRegulation* in figure 1(b). Notice that this time the column *targetSites* is not available, so he collects the *pValue* values. Also note that the scheme for

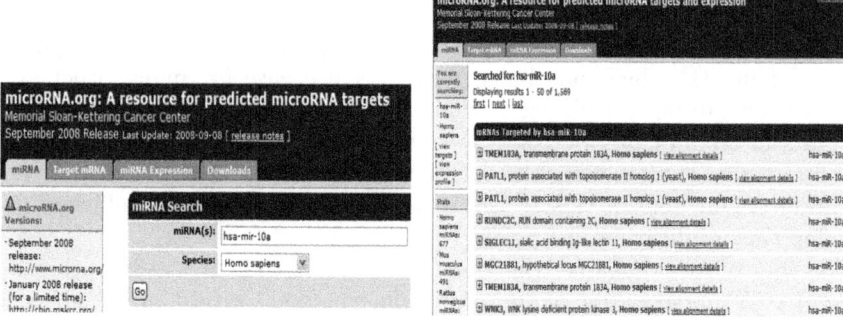

(a) microRNA.org input form.   (b) microRNA.org returned page.

Fig. 2. Typical user interaction interface at microRNA.org site

each of the tables is syntactically heterogeneous, but semantically similar (i.e., *miRNA* ≡ *microRNA, geneName* ≡ *geneID*, and so on). He does so because the data in the two databases are not identical, and there is a chance that querying only one site may not return all possible responses. Once these two tables are collected, he then takes a union of these two sets of gene names (in *micrornaRegulation* and *sangerRegulation*), and finally selects the genes from the intersection of the tables *proteinCodingGene* (that bind to p63, i.e., *p63Binding* = 'N') and *micrornaRegulation* ∪ *sangerRegulation* as his response.

To compute his answers in BioFlow using LifeDB, all he will need to do is execute the script shown in figure 3 that fully implements the application. In a recent work [48], we have shown how a visual interface called VizBuilder for BioFlow can be used to generate this script by an end user without any knowledge of BioFlow in a very short amount of time. It is also interesting to note that in this application, the total number of data manipulation statements used are only seven (statements numbered (2) through (8)). The rest of the statements are data definition statements needed in any solution using any other system. The meanings of these statements and the semantics of this script are discussed in detail in [23,54] and we refer readers to these documents for a more complete exposition. We would like to mention here that ad hoc integration, information aggregation, and workflow design using BioFlow in LifeDB is very simple compared to leading contemporary systems such as Taverna [49] and Kepler [14][1], and data management systems such as BioMediator [80], BioGuideSRS [27], and MetaQuerier [43][2]. Especially, the availability of graphical data integration and

---

[1] From our standpoint, Taverna stands at the same level as Kepler [14], Triana [70] and Pegasus [34] in terms of their architecture and operational semantics. We thus compare our system only with Taverna and Kepler in this paper for the sake of brevity.

[2] Except for BioMediator and an early system BioKleisli [33] and their descendants, most systems do not have a comprehensive data integration goal as does LifeDB, and as such the comparison is not truly even.

querying tools such as VizBuilder for BioFlow removes the need for actually writing applications making LifeDB a seriously appealing tool for Life Scientists. The algebraic foundations of these statements, called *Integra*, on the other hand, can be found in [46].

In the above script, the statements numbered (1) through (7) are most interesting and unique to BioFlow. The `define function` statements (2) and (4) essentially declare an interface to the web sites at URLs in the respective from clauses, i.e., `microrna.org` and `microrna.sanger.ac.uk`. The `extract` clause specifies what columns are of interest when the results of computation from the sites are available, whereas the `submit` clauses say what inputs need to be submitted. In these statements, it is not necessary that the users supply the exact variable names at the web site, or in the database. The wrapper (`FastWrap`) and the matcher (`OntoMatch`) named in the `using` clause and available in the named ontology `mirnaOntology`, actually establish the needed schema correspondence and the extraction rules needed to identify the results in the response page. Essentially, the `define function` statement acts as an interface between LifeDB and the web sites used in the applications. This statement was first introduced in [32] as the *remote user defined function* for databases where the input to the function is a set of tuples to which the function returns a table. However, the construct in [32] was too rigid and too mechanistic with the user needing to supply all the integration instructions. Actually, it could not use a wrapper or a schema matcher. The user needed to supply the exact scheme and exact data extraction rules. In BioFlow, it is now more declarative and intuitive.

To invoke the form functions and compute queries at these sites, we use `call` statements at (3) and (5). The first statement calls `getMiRNA` for every tuple in table `genes`, while the second call only sends one tuple to `getMiRNASanger` to collect the results in tables `micrornaRegulation` and `sangerRegulation`. The statements (6) and (7) are also new in BioFlow. They capture respectively the concepts of *vertical* and *horizontal* integration in the literature. The `combine` statement collects objects from multiple tables possibly having conflicting schemes into one table. To do so, it also uses a key identifier (such as `gordian` [76]) to recognize objects across tables. Such concepts have been investigated in the literature under the titles record linkage or object identification. For the purpose of this example, we adapted GORDIAN [76] as one of the key identifiers in BioFlow. The purpose of using a key identifier is to recognize the fields in the constituent relations that essentially make up the object key[3], so that we can avoid collecting non-unique objects in the result. The `link` statement, on the other hand, extends an object in a way similar to join operation in relational algebra. Here too, the schema matcher and the key identifier play an important role. Finally, the whole script can be stored as a named *process* and reused using BioFlow's `perform` statement. In this example, line (1) shows that this process is named `compute_mirna` and can be stored as such for later use.

---

[3] Note that object key in this case is not necessarily the primary keys of the participating relations.

```
process compute_mirna                                                    (1)
{ open database bioflow_mirna;
  drop table if exists genes;
  create datatable genes {
    chromosome varchar(20), start int, end int, miRNA varchar(20) };
  load data local infile '/genes.txt'
    into table genes fields terminated by '\t'
    lines terminated by '\r\n';
  drop table if exists proteinCodingGene;
  create datatable proteinCodingGene {
    Gene varchar(200), p63binding varchar(20) };
  load data local infile '/proteinCodingGene.txt'
    into table proteinCodingGenes fields terminated by '\t'
    lines terminated by '\r\n';
  drop table if exists micrornaRegulation;
  create datatable micrornaRegulation {
    mirna varchar(200), targetsites varchar(200), geneID varchar(300) };
  define function getMiRNA
    extract mirna varchar(100), targetsites varchar(200),
    geneID varchar(300)
    using wrapper mirnaWrapper in ontology mirnaOntology
    from "http://www.microrna.org/microrna/getTargets.do"
    submit( matureName varchar(100), organism varchar(300) );        (2)
  insert into micrornaRegulation
    call getMiRNA select miRNA, '9606' from genes ;                  (3)
  drop table if exists sangerRegulation;
  create datatable sangerRegulation {
    microRNA varchar(200), geneName varchar(200), pvalue varchar(200) };
  define function getMiRNASanger
    extract microRNA varchar(200), geneName varchar(200),
    pvalue varchar(30)
    using wrapper mirnaWrapper in ontology mirnaOntology
    from "http://microrna.sanger.ac.uk/cgi-bin/targets/v5/hit_list.pl/"
    submit(  mirna_id varchar(300), genome_id varchar(100) );        (4)
  insert into sangerRegulation
    call getMiRNASanger select miRNA, '2964' from genes ;           (5)
  create view regulation as
  combine micrornaRegulation, sangerRegulation
    using matcher OntoMatch identifier gordian;                      (6)
  create view proteinCodingGeneRegulation as
  link regulation, proteinCodingGene
    using matcher OntoMatch identifier gordian;                      (7)
  select *
    from proteinCodingGeneRegulation
    where pValue <= 0.006 and targetSites >= 2 and p63binding='N';    (8)
  close database bioflow_mirna; }
```

Fig. 3. BioFlow script implementing the process

3.1   Related Research

Before we present the highlights of LifeDB system, we would like to mention where it stands relative to its predecessors. There are several well known data integration systems in the literature. We single out only a few more recent ones, simply because they improve upon many older ones, and are close to LifeDB. BioKleisli [33] and Biopipe [45] are two such data integration systems for life sciences that help define workflows and execute queries. While they are useful, they actually fall into the second category of application design approaches discussed in section 1. As such, they are very tightly coupled with the source databases because they directly interact with them using the services supported by the sources. The scripting language just eases the tediousness of developing applications. Schema mediation remains user responsibility and they do not deal with hidden web resources, or ad hoc data integration. One of the systems, ALADIN [21] (ALmost Automatic Data INtegration), falls into the first category and supports integration by locally creating a physical relational database warehouse manually. However, it uses technologies such as key identification and record linkage to mediate schema heterogeneity, applying domain and application specific knowledge and thus having limited application.

In LifeDB, integration is declarative, fully automatic and does not rely on a local copy of the resources (it uses virtual warehousing). Neither does it depend on application or domain specific knowledge. For all queries, workflows and data integration requests, the schema mediation, wrapper generation and information extraction are carried out in real time. One of the recent integration systems, MetaQuerier [43], has features similar to LifeDB. It integrates hidden web sources, and uses components similar to LifeDB. But MetaQuerier integrates the components using glue codes, making it resistant to change. In LifeDB on the other hand, we focus on a framework where no code writing will be necessary for the development of any application regardless of the sites or resources used. In addition, the application will never require specialized site cooperation to function.

LifeDB stands out among all the systems in two principal ways. First, it extends SQL with automatic schema mediation, horizontal and vertical integration, hidden web access, and process definition primitives. Apart from the fact that for the purpose of programming convenience, it retains a few non-declarative language constructs such as assignment and loop statements, it is almost entirely declarative. It separates system aspects from the language, making it possible to change the underlying system components as newer and better technologies become available, and improves its functionality without compromising the semantics of its query language, BioFlow. Second, it successfully avoids physical warehousing without sacrificing autonomy of the sources – in other words, it does not depend on specialized services from the source databases such as option to run stored scripts, access to the server to submit scripts for execution and so on, the way BioKleisli or Biopipe require. Yet, it supports declarative integration at a throw away cost. Unlike ALADIN and MetaQuerier, it is also fully automatic, and no user intervention or glue code writing is required to enable integration and access.

3.2   Novelty of BioFlow over Web Services and Workflow Engines

It is possible to make a case for an approach based on emerging technologies such as XML, semantic web or web services. In many recent research (e.g., [67]), arguments have been made that adopting such technologies makes it possible for an improved service composition and workflow development experience as the one discussed in the earlier section. In fact, when web service based systems fully develop and all the providers of life sciences resources adopt this technology and adhere to standards such as SOAP, WSDL and so on, researchers will have a much better platform to develop their applications. Unfortunately, we estimate that the time line for such a scenario is at least several years in the future. This assertion holds because there are more than several thousand life sciences databases and resources on the internet that biologists use. Only a handful of these databases are web service compliant, i.e., *maxdLoad2* database [42]. Only recently, PDB has begun to redesign its web site and applications in web service compliant formats with an aim to better serve its users that is expected to take some time to complete and stabilize [86]. Platform migration (to web services) needs significant motivation, huge resources, and provider incentives. Even new applications that emerge sporadically are based on simple, quick and dirty conventional platforms.

We argue that in the interim, we need to develop middleware that are equally capable of exploiting the conventional or legacy systems as well as more advanced and emerging systems based on web service type technologies. This was exactly the argument behind developing LifeDB and we have demonstrated in section 3 that LifeDB is capable of delivering that promise. What is not apparent in the discussion in section 3 is that systems that rely on emerging technologies solely, fail to exploit almost all the life sciences resources that currently exist, which have been developed at a huge expense and are likely to continue to support the conventional platform for an indefinite period. As we have argued before that current trend has been to reuse resources and not warehouse them. So, it is not likely that users will choose to modernize resources on their own by replicating them to fit their application.

As a demonstration of BioFlow's suitability over systems developed solely based on web service type technologies, we have objectively compared BioFlow with leading workflow systems such as Taverna and Kepler in [54] using practical examples. These microarray data analysis examples have been described as fairly complicated workflows [67] requiring the sophistication and modeling power of leading workflow systems such as Taverna and Kepler. Taverna is a leading workflow management system developed in the context of [my]Grid project, and is widely favored in life sciences workflow application design. We have reproduced and adapted the workflow in [67] in LifeDB using BioFlow. The BioFlow implementation of the same workflow, as presented, can be written fairly fast by end users, and can be developed using BioFlow's front end VizBuilder [48] literally in minutes. It is again important to note that the BioFlow implementation does not require any code writing, specific low level knowledge, or site cooperation of any form from *maxdLoad2* database, in this instance.

We also argue that application design using Taverna for applications similar to the one in section 3 will be even more involved. This is particularly so because the *microRNA.org* and *sanger.org* do not have web services, or all the services required for Taverna to operate. So, an application designer will have to necessarily write all communication and data manipulation codes to implement the workflow and develop all schema mediation mappings manually[4]. Furthermore, while Taverna is a mature system, it still does not provide any support for schema heterogeneity mediation, whereas in BioFlow, no code writing[5] is required whatsoever for most applications. In fact, the script presented in section 3 was developed again using VizBuilder in minutes. An example session of BioFlow application design is shown in figure 4.

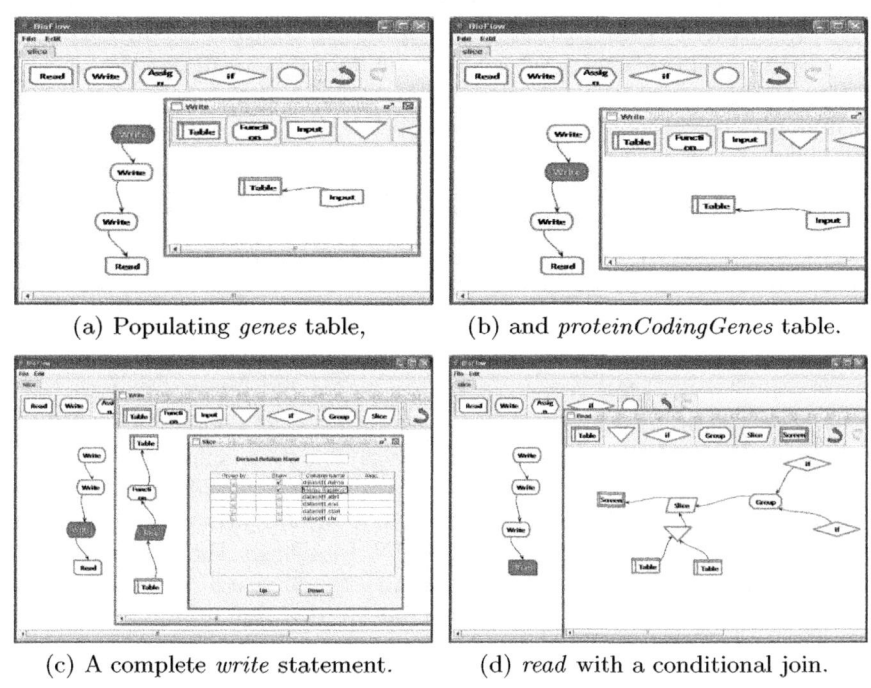

(a) Populating *genes* table,   (b) and *proteinCodingGenes* table.

(c) A complete *write* statement.   (d) *read* with a conditional join.

Fig. 4. Partial miRNA application script development using VizBuilder

---

[4] It is interesting to note here that the maxdLoad2 database and the Taverna systems were developed by two groups in the Computer Science Department, University of Manchester, and hence there is little evidence that even for Taverna, web services delivered the promise it holds.

[5] We contrast code writing as in Java, Perl, PHP or C with ad hoc querying in SQL like declarative languages by arguing that declarative languages are more abstract and conceptual in nature, and not procedural.

3.3   Why Choose LifeDB

We believe LifeDB offers three distinct advantages over traditional systems and other emerging systems based on web services. First, it offers higher level abstractions for scientific workflow development with support for data integration which biologists desperately need. These abstractions are easy to cast into SQL like languages (such as BioFlow) and then develop visual interfaces (such as VizBuilder) to help develop applications on an ad hoc basis at throw away cost. Second, it supports current and future trends in service integration to help systems biology research. It does not encourage warehousing and expensive glue code writing or programming, and helps migration to different platforms truly effortlessly. Once a prototype application is developed using distributed resources, it is easy to materialize the resources needed locally yet use the same application should network latency become an issue for high volume data processing, a relatively rare scenario. Finally, until web service or semantic web based systems become widely available and all popular life sciences resources migrate to such platforms, LifeDB is suitable as a transitional middleware platform. The advantage it has is that all the applications developed using LifeDB will still function in modern platforms such as web services, as demonstrated in the example in section 3.2 and in [54].

4   Limitations of BioFlow and New Challenges

Since our focus in this paper is to introduce LifeDB and BioFlow on intuitive grounds, highlight their unique strengths and raise new challenges that are needed to be addressed to make LifeDB, or any other system that addresses autonomous and ad hoc data integration, a viable data management platform, we invite readers to LifeDB related articles in [23,46,54,39,24,19] for a comprehensive exposition. In this section, we assume readers' familiarity with LifeDB, its data management capabilities and underlying tools such as FastWrap and OntoMatch LifeDB uses to execute BioFlow queries. It should not be difficult to see that although LifeDB and BioFlow have many features that set them apart from contemporary systems such as ALADIN, BioKleisly, BioGuideSRS and others, and offer a significantly novel model, they too have a number of limitations. For example, although LifeDB addressed data integration from the viewpoints of schema heterogeneity and distributivity of resources, and information aggregation, it did so for data types that are flat (traditional relational data) and crisp (error-free and certain data). As biologists have started to appreciate the data integration support and convenient data management support of LifeDB, it is becoming increasingly apparent that without the support for complex data types, improved workflow development tools, and management and reasoning tools for non-crisp data, its true potential may remain unexploited. In the absence of such supports, even LifeDB users currently resort to customized application and tool development that is time consuming, error prone, and expensive. Our goal in this section is to identify the major shortcomings of LifeDB as new challenges that once resolved, will strengthen the data integration model it advocates and set a new standard for distributed data management. We identify these limitations in two distinct axes: (i) data model and query language dimension (discussed in section 5), and (ii) core query processing engine artifacts (discussed in section 6). In the next few sections, we present these challenges in the light of user experience, and what researchers, including our lab, have already investigated to address some of these issues. We hope to convince readers that these challenges are novel and interesting and that there are ways to solving them.

Before we discuss these research issues, let us consider another example query taken from [13] to illustrate that life sciences applications do not always deal with flat tables as in the microRNA example in section 3 and often need manipulation of very complex and emerging data types for which tools are not often available, forcing custom application development. The query *"Prioritize all the genes in the human proteome for disease association based on human-mouse conserved co-expression analysis"*, essentially requires access to and manipulation of various kinds of data types such as expression, interaction network, and flat tables. The steps below describe one possible workflow to implement this query, which can be easily done in BioFlow (actually the BioFlow script and the implementation is available online at LifeDB home page [7]).

1. Go to Stanford Microarray Database (SMD).
2. Download 4129 experiments for 102296 EST probes for human and 467 experiments for 80595 EST probes for mouse.
3. Use Homologene to create the conserved interaction network between human and mouse. Call it 'Stanford' network.
4. Go to Affymatrix Microarray Database
5. Download 353 experiments corresponding to 65 different tissues for 46241 probe-sets associated to a known gene for human and 122 experiments corresponding to 61 tissues for 19692 probe-sets for mouse.
6. Use Homologene to create the conserved interaction network between human and mouse. Call it Affy network.
7. Merge 'Stanford' and 'Affy' networks and call it PPINet.
8. Go to OMIM Morbid Map and download/search loci information for the genes in the network.
9. For each gene
   (a) Search all the interacting partners of the gene from the PPINet.
   (b) List the OMIM IDs associated to the interactors from the OMIM Morbid Map Loci Info table.
   (c) Go to MimMiner (http://www.cmbi.ru.nl/MimMiner/cgi-bin/main.pl). For each pair of the OMIM IDs calculate the phenotype similarity score by using MimMiner.
   (d) Compute the average score obtained from previous step.
10. Sort the list of genes according to the computed scores from the previous step.

In the above example, not only do users access and manipulate flat tables, they also need to analyze interaction data (graphs), and gene expression data (vector type). Fortunately for this example, multiple online analysis tools exist that have been integrated into a LifeDB application. However, if the response returned by these tools are in complex form, we will have no choice but to treat each one of them as separate applications. The current approach for integration of these data types in LifeDB is to apply off line tools in a decoupled fashion and import the results to LifeDB applications in a suitable input form. In contrast to the example above, the workflow diagram shown in figure 5 points to a more system level hurdle that needs serious attention.

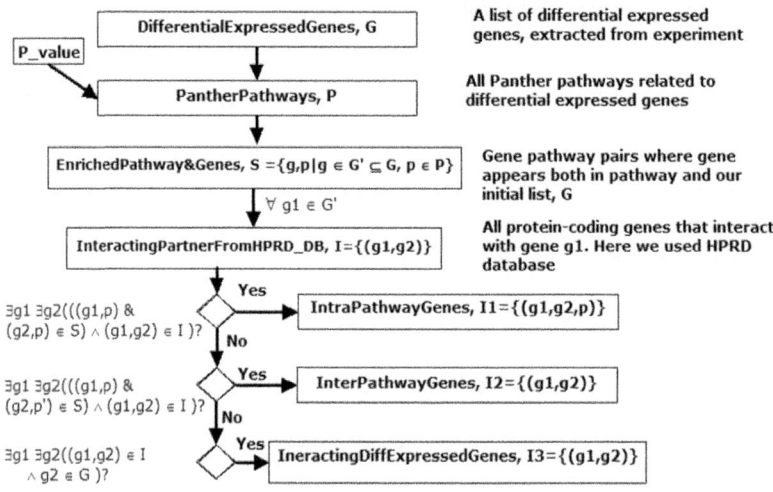

Fig. 5. Human-mouse conserved co-expression analysis pipeline

In this example, a user has created a custom set of a gene expression for an epilepsy study from human brain cells. His goal is to find relationships among genes that are expressed in the study with genes found in known pathways, and protein interaction databases. The goal is to study gene relationships before a hypothesis is formed and tested. To find these relationships, the user chooses to use the Panther database for gene expression analysis, and HPRD database for protein interaction study. The various conditions and decision steps are depicted in the workflow diagram showing how the results are generated.

Although development of an application to execute this analysis is easily available in LifeDB, the application can not be fully automated. The reasons are not truly related to LifeDB's ability, but more to its component technologies failed to perform their intended functions. These failures, in turn, are intimately related to how newer web technologies based on Java Script and Ajax are shaping web application development practices. Since we choose to access and use online resources as they are and adopt web page scrapping for the implementation of our wrapper system FastWrap, these technologies have a serious impact on our system. We will revisit this issue in section 6.1 in greater detail.

5   Data Model and Query Language Enhancement

From the standpoint of data model and query language, we identify three major research issues – (i) XML-Relational mixed mode operation, (ii) graph and expression data management, and (iii) uncertain or noisy data management. We elaborate on these issues next.

5.1   XML-Relational Dichotomy

From the standpoint of end users in Biology, flat view of data is perhaps the most acceptable of all formats while some form of shallow nesting is not truly difficult. Therefore, it can be argued that the relational data model fits well with the practice and end user psychology well. While XML has been steadily increasing in popularity, querying such data sets using languages such as XQuery or XPath is considered truly difficult still. As argued in works such as BioBike [35], for over several decades, biologists have voted overwhelmingly not to embrace computer programming as a basic tool through which to look at their world, leading to the field using natural language processing (NLP) as a means to gain access to needed information from the vast array of life sciences repositories. In reality, however, such a high level interface still remains illusive, mostly because of the representation hurdles, highly interpretive nature of life sciences data[6], translation from natural language to database queries, and the inherent difficulty in processing NLP queries. A practical approach, we argue, is providing a flat relational view of complex data such as XML so that users can comprehend and query the information repositories at their disposal in the well understood model of flat relations as a middle ground.

Unfortunately, many biological databases today export data in complex formats other than flat tables including plain text and XML. Databases such as (i) BioGrid, HPRD, and DroID [83] (interaction and pathway data); (ii) PDB, OCA [9] and ModBase (protein structure data); (iii) GO and TOPSAN [10] (annotations); (iv) GenBank and FlyBase (sequences); (v) TreeFAM [66], PhyloFinder [31], TreeBASE (phylogentic data), and (vi) Geo and ArrayExpress (expression data) export data in various formats including many exchange standards that are derivatives of XML such as MIAME, MAGE-ML, HUPO, ProML, KGML, PDBML, NexML, PhyloXML, etc. Consequently, it is imperative that we accommodate representation and processing of data in these formats. Although LifeDB adopts a virtual warehousing approach (no materialization of views), it still needs to process the data returned by the application sites that are potentially in mixed formats. Recent work such as [84] adequately demonstrates that even though standards have been adopted, its use and interpretation is still a case by case approach and the responsibility of the application designers.

Traditional approaches to mixed-mode data processing involving XML and relations include query transformation (e.g., [36,50]), and model translation (e.g.,

---

[6] Tool applications are essential before an understanding can be gained for most biological data such as DNA sequences, protein structure, or pathways. Simple read off of the data as in relational model does not reveal any information in general.

[26,30]). In this approach, the data from one model is physically transformed into another for a uniform user view, and queries are then transformed to match the user view of data. In stable environments, such transformation techniques may yield modeling benefits and allow ease of query processing. But in loosely integrated and highly volatile databases, the likely gain in modeling ease is often nullified by excessive query processing time due to back and forth query and data translation. Thus, in an environment such as LifeDB where response time is already high due to network latency and high volume of data transmission, every bit of additional processing time leads to degrading overall performance. Hence it is desirable to formulate an approach in which too much transformation is not warranted, and a flat view of data, which biologists prefer, is maintained.

In an attempt to offer such an alternative, we have considered two principal approaches, neither of which could yet be adopted due to inadequate performance results and research. In both approaches, the database content is never transformed and is maintained in its original format. For the first approach, we have considered the idea of *Information Content Tableaux* (ICT) [60] in which the view of XML documents is changed and made relational in which:

- using a notion of valuations from an XML database schema to the XML database, we capture the latter's information content, develop an alternative computational semantics for the information content and show it is equivalent to the valuation-based notion of information content, and show that the information content can be used to faithfully reconstruct the original XML database.
- the information content can be represented as a relational database, which offers a high level information-preserving view of the original XML database. With this user interface, the user can write queries in SQL which can be translated into equivalent XQuery queries against the original XML database (i.e., [50]).
- to support flexible XML output construction, a lightweight extension to SQL, called SQL-X, has also been proposed [60]. This translation algorithm can correctly translate queries expressed in SQL-X[7], and was shown to capture a rich class of XQuery queries using the information content view as a front end and SQL(-X) as a query language.

In the second approach, we simply rewrite SQL/BioFlow queries to XQuery to execute against the XML databases, and integrate XML-relational model via reconciliation. In other words, we develop a new paradigm for query inter-operation wherein we envision a single query to range over native relations and XML documents. In such a scenario, we propose that a high level nested relational view of data be maintained while the lower level data is in either relational model or in XML. In fact, in BioFlow, we have included syntax that hides the relational and XML dichotomy of table formats by viewing them uniformly. The create datatable declaration recognizes the difference, yet the select from where statement does not make any distinction between XML and relational data. However,

---

[7] It should be mentioned here that SQL/XML [57,38] has similar capabilities and can be used as an alternative without any loss of the developments we propose.

in the current implementation we do not allow mixed mode processing because technical hurdles remain. This issue is further complicated by the fact that in BioFlow we allow user applications to include a transparent mix of databases such as MySQL, Oracle, DB2, and any native XML database such as MonetDB.

5.2   Integrated Graph and Tree Processing

Experimental methods are beginning to define the networks of interacting genes and proteins that control most biological processes. Scientific interpretation of such complex networks, generally called biological networks, require sophisticated algorithms and tools. They can be classified into protein interaction networks (protein-protein, protein-gene, gene-gene, protein-RNA, etc.), structures (protein and molecules), and pathways (metabolic and signaling). Another type of network data, called phylogenetic data, captures evolutionary relationships of species and molecules. Although these types of data are a special class of graph like data, they stand on their own in terms of their use, complexity, analysis and uniqueness in Biology. As demonstrated earlier, in systems biology, the interrelationship among all these types of data is deep, intricate and complex. In the previous section, we have already introduced some of the leading databases for each category. But to our knowledge, no native database query processing capabilities have been implemented in any leading data management system such as Oracle, DB2, MySQL or SQL Server which these application databases could leverage. Consequently, analysis tools are the common vehicle for interpreting and querying graph-like data.

This separation creates application development hurdles in which users are forced to switch between multiple platforms, tools, and formats, creating a data management nightmare because of version management and identification related complexities. Although in LifeDB we support fairly powerful yet simple tool integration primitives, it does not always work out well. For example, data needs to be pre- and post processed before and after tool application to integrate with other database data, and often these tools require manual intervention or read off of data from graphical displays. We believe graph-like data should be part of a data management suit that users can rely on for most of their query processing needs, and format disparities should be handled in the background without user involvement, if possible. In that direction, we think beyond commonly used graph querying functions such as lookup, match, isomorph, and so on, other advanced functions can be added as functions that users can just call up as part of a query.

Since the co-habitation of XML and relational data is here to stay, and we are already dealing with both in LifeDB, it is essentially immaterial which basic model one might want to follow to model graphs at the database level. From this standpoint, we believe two contemporary research directions appear interesting. The first one is a class of systems that blend data model and query language to extend basic graph processing support. The representative query language we find most appealing is GraphQL [44]. This langauge supports arbitrary attributes on nodes, edges, and graphs and graphs are treated as first class citizens, and the model allows querying graphs as basic sets of units. The language is supported by typical select, project and join type algebraic operators and query processing plans that build upon relational algebra. The join operation exploits flexibilities supported in the model for graph structure composition. The selection operator has been generalized to graph pattern matching and a composition operator is introduced for rewriting matched graphs. It was shown that the efficiency in processing queries is acceptable.

Another promising direction is a suit of algorithms that do not require exotic data structures to start with in order to provide specialized query support and could leverage graphs represented using traditional data models such as relations. In our own laboratory, we are developing a set of graph algorithms to aid network analysis and querying. As part of of our Network Analysis Query Language (NyQL) project, we have developed a graph representation model and query processing technique based on traditional relational model [51,52]. The novelty of this model is that we can process arbitrarily large graphs that are possibly disk resident. Yet, we have shown that our algorithm performs better than leading algorithms such as VFLib [79] and Ullmann [78]. The representation makes it possible to search, query or match graphs with arbitrary conditions using isomorphism as the basic backbone. In fact the works in [51,52] has been implemented in logic programming systems such as XSB [11] and Prolog, as well as in MySQL. Evidently, the queries are first-order expressible, and since it is declarative, optimization opportunities can be explored as an orthogonal research. The applicability of this approach has been demonstrated in our most recent works on extending Cytoscape and KEGG pathway database [77]. We feel strongly that with some adaptation, we can also use our algorithm for network alignment [81,40] which is now in the forefront of pathway and interaction network research in Biology. Finally, our own PhyQL system [47] was also implemented in XSB [11] without the requirement of any tree representation scheme used in many contemporary research. With the help of the proposed graph technique, we can now implement tree queries without restrictions and more efficiently. Since the algorithm we have developed can be expressed declaratively in logic or on SQL, it is truly possible to develop a query language that will be declarative as well.

Since not all types of queries will be expressible in NyQL, we plan to develop a set of algorithms such as graph matching [15], and alignment [25]. These can then be combined with querying techniques already introduced in [51,52] that can be used as graph functions toward applications such as disease gene identification [17]. Regardless of how new algorithms in functions work, these functions should be available to the query language for data processing in their native format. Only then can we hope to remove the impedance mismatch and platform disparity we seek.

5.3   Uncertain and Conflicting Data Management

Noise, uncertainty and error are the facts of life Biologists live with everyday. In the absence of a rigorous mathematical basis on which data processing and analysis tools can be built, the practice is to assume data to be correct and deal with faulty results by assigning an object credibility. Often these accuracy indicators conflict and point to multiple true positives when there should be only one. Usually there is no option for users to back analyze the data to find out the source of the error, and thus there is no systematic way of error correction. There is a wealth of formal query languages and data models capable of handling uncertain data such as imprecise data, uncertain and probabilistic data, fuzzy and rough sets, may be or null values, and so on. But the complex nature of biological data does not fit into many of these models and hence, was essentially ignored. Biologists themselves did not make a concerted effort to generate such data. For example, although high throughput protein interaction data is known to be imprecise and noisy, not much has been done to quantify the error as part of basic data sets. Instead, the quality of the lab as a whole has been assumed to be an indicator of the quality of the data. Only recently, as part of the DroID [83] database, Drosophila protein interaction data from numerous sources has been collected. These data were noisy in the sense that conflicts, errors and uncertainties were a major attribute of the data and its collection process. Several inconsistency resolution models and curation technique have been applied before including the data in this database. The data is of such good quality that DroID is now part of FlyBase and these two databases are mutually cross linked. The details of the integration process may be found in the DroID home page [83] and in [82]. There are several other parallel efforts that aim to attribute such quality scores to the data themselves. Perhaps, these emerging sets of data will encourage researchers to take a fresh look at uncertain complex data processing and propose solutions.

In this context it is probably interesting to consider the *Information Source Tracking* method (IST) [75,59] to manage uncertainty of data. This model allows manipulation of information sources to assign reliability in a way similar to current practices of assigning trust in the quality of the data based on the lab. IST has been later extended to include parametrization [62] to make it a general framework for uncertainty management to support fuzzy logic, probability, rough sets, etc. in the same platform. The advantage of this model is that reliability has already been made part of the information source, and a complete data model and query language has been developed. We need to make necessary adjustments to make it suitable for life sciences data. However, the complexity of handling uncertain data is usually very high, and given the size of biological data sets, the practicality of this endeavor is still an open question.

5.4   Supporting Gene Expression Data

Expression data constitutes a significant portion of biological resources today. Strikingly, not much attention has been devoted to the management and querying of these data so far. The main focus has been on developing application tools to gain access and understand the collections in all main expression data repositories. This is partly because the data is complex, needs several layers of cleaning, preprocessing and orchestrating before application tools can be applied to unravel their information content because traditional data management tools cannot be applied directly. The standards such as MIAME [28] and MAGE-ML [8] are put in place to help digest the data in an understandable format than in their native expression vector form emerging from hybridization technologies inherent in traditional Affymetrix [2], Solexa [22], Illumina [6] and Agilent [3] chips.

In principle, gene expression data can be viewed as providing just the three-valued expression profiles of target biological elements relative to an experiment at hand. Although complicated, gathering the expression profiles does not pose much of a challenge from a query language standpoint. What is interesting is how these expression profiles are used to tease out information from the vast array of information repositories that associate meaning to the expression profiles. Since such annotations are inherently experiment specific functions, much the same way queries in databases are, developing a querying system for gene expression data appears to be pointless. Instead, developing tools and techniques to support individual assignment has been considered prudent.

In our recent effort to help systems biology studies, we proposed a new platform independent and general purpose declarative query language called *Curray* [53], for Custom Microarray query language, to support online expression data analysis using distributed resources as sub language for BioFlow to nimbly allow flexible modeling and room for customization. In Curray, to view all expression data uniformly, we make a clear separation between the conceptual components of such data and their platform (technology, class and type) related components so that at the application level, queries can be asked in uniform ways without referring to low level details. To offer control of data to the user, we also allow drill down and roll up type of concepts on expression data. For example, using R/Bioconductor's affy library [1] the probe set intensities can be summarized to form one expression value for each gene.

From figure 6, it will be evident that Curray allows incorporation of library functions from packages such as Bioconductor/R [73] and statistical packages such as SPSS and SAS. Our experience show that such a marriage is barely workable as it compromises the declarative nature of the language and moves fast toward a procedural language such as Perl or BioPerl the biologists already do not like much. Further research is needed to make Curray a truly declarative language capable of supporting user specific expression data analysis without having to write application codes.

It is easy to see in figure 6 that complex analysis that is not covered by the basic Curray statements can be easily performed using BioFlow. The most enabling feature in Curray is the way expression data is interpreted in the create expressiontable and create expressionview statements, and how they are organized behind the scene so that users need not worry about the details. Furthermore, the way an SQL-like feeling is offered is novel, wherein analysis functions are blended through the using library option in the having clause. It is however fair to say that although the basic Curray expression functions follow the suggestions in [72] that advocates four basic functionalities – class discovery, class comparison,

extract into *epilepsyData*   compute *regulation, . . .*
from *epilepsyDataLims*   from rollup *epilepsyDataLims* to *expression*
where *θ*   using function *functionName*
using function *functionName*; where *θ*;
          (a)   (b)

define function *davidInput* URL
from *http://david.abcc.ncifcrf.gov/summary.jsp*
submit (*geneList* string, *typeOfID* string);

define function *david*   extract into *epilepsyMageml*
extract *pathway* string, *relatedGenes* string, *fdr* float   from *epilepsyData*
using wrapper *davidWrapper* mapper *davidMapper*   where *θ*
from URL *davidUrl*   using function *functionName*;
submit (*species* string | *davidUrl* string);   (d)

call *david* ('homo sapiens',
  call *davidInput* with *diffExpressGeneList*);
          (c)

Fig. 6. Data definition and data manipulation statements in Curray

class prediction and mechanistic studies (figure 6 depicts a few representative statements), there are numerous other types of analysis needs that are pretty common in this domain that could also be supported.

6 Query Engine Improvement

LifeDB query processing engine is a confluence of several sub-systems such as back end database management systems MonetDB [85], MySQL, eXist [5], wrapper generation system FastWrap, schema matching system OntoMatch, visual query engine VizBuilder, and entity resolution system Gordian. The performance and overall quality of BioFlow depend largely upon the underlying quality of its sub components that were developed independently. In this section, our goal is to identify two critical shortcomings that are greatly impacting LifeDB in particular, and online data processing in service composition in general. We believe addressing these issues is important in their own right, independent of LifeDB. The first research issue is related to wrapper generation in general, and the second is related to user specified optimization at the server side query processing.

6.1 Wrapper Generation for Emerging Applications

The casting of Integra in BioFlow, and its implementation required the development of a fully automatic and efficient wrapper generator FastWrap based on web scraping, and a schema matcher OntoMatch having features suitable for BioFlow. The **define function** statement (the *transform* operator *τ*) uses the services of FastWrap to extract a target table from the HTML documents returned by a website. We used a suffix tree based implementation that has linear time complexity and hence, online performance is not an issue. Once extracted, the scheme of the table is matched with the user query variables to resolve schema heterogeneity at run time. Our OntoMatch system has a quality guarantee that many systems do not – the match quality in OntoMatch improves monotonically as successive lower level matchers are applied, and most erratic behavior of contemporary matchers are avoided. This is an essential feature for an unsupervised automated run time matcher where user intervention is not expected. While the wrapper and matcher work elegantly, some website responses pose a serious problem when they do not annotate the tables they return as responses, and FastWrap fails to supply a fully annotated table to OntoMatch for schema matching. They also create problems for an automated wrapper system when newer technologies such as Java Scripts are used for multi-page displays.

Column Name Annotation. Most websites are designed for human use and hence, a certain degree of user interaction is expected. As a consequence, many sites do not annotate the data they provide as a response with attribute names because they are usually evident from the context, for a human agent. The current solution in BioFlow has been to throw the extracted but unannotated table at the user for identifying the columns manually. But once annotated, it is stored in the system until the site changes. Although the adopted solution works in the event no attribute names are found, it is not fully automatic, defeats the purpose of BioFlow, and imposes significant burden on the user. To alleviate this unwanted situation, we have experimented with three different fully automatic annotation techniques.

The first annotator is based on grammar rules for attribute values [18]. The method depends upon maintaining an ontology of rules for different types of data, even domain dependent data, and letting the ontology enrich itself by learning new rules as queries are processed. It proceeds as follows. Data table, once extracted, is subjected to a dual scanning mechanism i.e. horizontal and vertical. The vertical scan is performed for each column, where we focus on understanding the structural commonality that may exist among the data in the same column. The structural information is expressed in regular expression in terms of numeric characters, special characters, and string of alphabets.

As a first step of the processing, we initially iteratively process data pertaining to a single column to identify hidden structural patterns associated with it. We first extract the largest common prefix and largest common suffix pertaining to all values in the column vector. If so, we retain those values since they are likely to provide semantic information. Next, we split each and every entry in the column vector in terms of constituent parts of structural components and create a generic regular expression corresponding to that entry. However, it is likely that not all of the entries in a column vector will adhere to the same structural pattern. Hence, we combine all the entries by creating a global regular expression that can satisfy all the elements in the column vector. This is done by vertically aligning the structures using multiple sequence alignment techniques.

While the above approach works for strings having patterns of different types, value strings that do not have any variability in string types (character, special character, integer, etc.), are hard to extract type rules. For these types of values, we have developed a new complementary technique for annotation based on Wikipedia [16]. Since Wikipedia provides a comprehensive repository of semantic and structural relationships among concepts, it can be regarded as an ideal knowledgebase for missing column name annotation purposes.

Wikipedia contains a mapping between concepts and their inherent categories as perceived by the community. Moreover, in order to cross reference categories, Wikipedia also provides a category table that delineates the relationship between multiple categories i.e. whether category A is a specialization of Category B or vice versa. Thus, the entire category table can be viewed as a graph where the nodes represent categories and the edges represent semantic relationship among the nodes. We can extend this graph by adding directed edges from categories to the respective concepts. This categorization serves as the annotation of the columns. While this method works better than the first, it is extremely slow and not suitable for online data integration.

Finally, often there may be situations where neither techniques are useful. In [12], we have exploited the power of the internet documents to efficiently annotate objects in a given column using search engines such as Google, Yahoo and Bing. In this approach, label value pairs $(< L, V >)$ are searched on the net and the counts of such co-occurrences are collected with the assumption that for any pair of terms, the term on the left is the annotation and the other is its value. For this approach to work, we must have a good understanding of the domain; for that domain, we consult an authoritative site (such as UIUC BAMM database [4]) for candidate labels and use those labels to formulate "speculative labels" in the form of label value pairs and submit to a set of search engine for hit count. We can use a sampling method to choose a very small but representative subset of values in the un-annotated table to restrict computational cost for the speculative labels. Once the hit counts are collected, we can use standard statistical methods to compute the annotations. The advantage of this method is that it is relatively faster, but the disadvantage is that often the quality of annotation is poor due to overlapping column domains, and tend to be be slower for larger tables. The method is also not suitable for tables with values that are not part of well understood textual information (e.g., model THG55 Panasonic TV), or when we are unable to supply a well understood set of candidate labels to the annotators to select from.

Java Script Enabled and Other Exotic Displays. Increasingly developers are choosing to use Ajax type Java Scripts for page design with greater control and dynamic generation of content from server side data repositories. They are also being used data validation, local processing and security purposes. Web scrapping technology used in FastWrap, or any other wrapper generation system, is not always capable of dealing with Java Scripts. This issue may be better explained with figures 7 through 9, all of which are using Java Scripts. These are the response pages generated by HPRD/BioGRID database as part of the

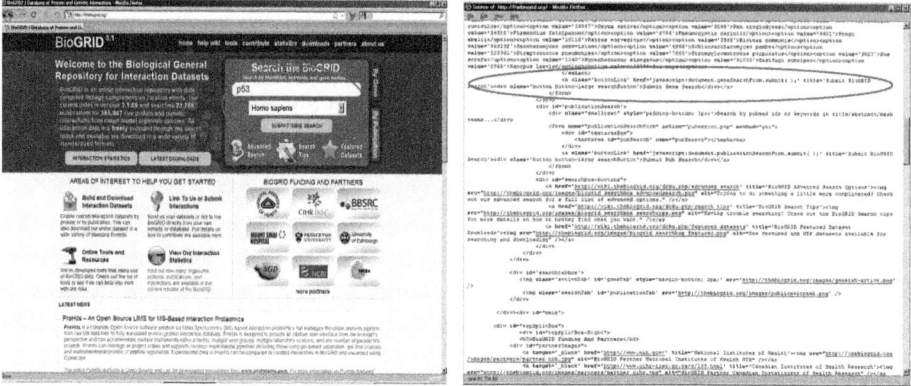

(a) Visible submission form.   (b) Form submission script.

Fig. 7. This form is being submitted using a script, and hence FastWrap cannot see what to submit during scrapping

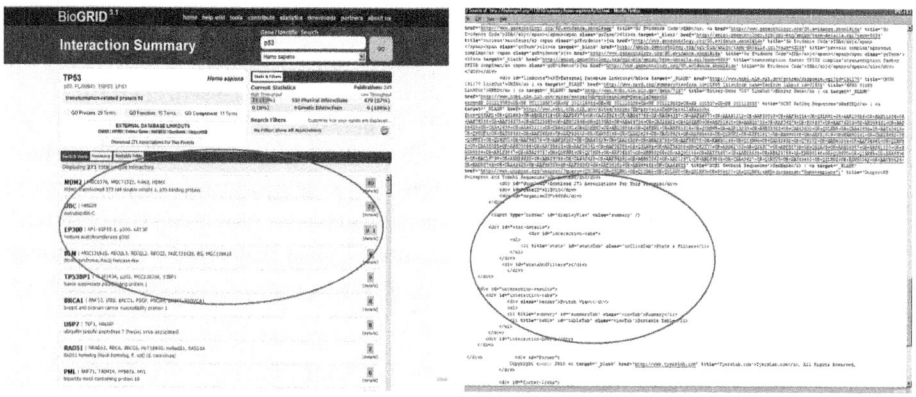

(a) Table not part of the document.   (b) The actual script visible to FastWrap.

Fig. 8. FastWrap fails to decipher and extract the Java Script in figure 8(b) used to display the table in figure 8(a)

workflow shown in figure 5 which could not be fully automated due to the presence of Java Scripts.

In figure 7(a), we show what is visible to the user but the submit button is not implemented using the POST method as is done in traditional web pages. Figure 7(b) shows the script of this page that led to the failure of our wrapper FastWrap. Similar problem appears when tables are created using Ajax dynamically as shown in figures 8(a) and 8(b). In this instance, our wrapper could not identify or extract the table.

Finally, multi-page tables are hard to recognize as scripts determine how to navigate to the next page, not the next button as shown in figure 9(a), and tables that have less repetitive patterns are equally hard to find (shown in figure 9(b)).

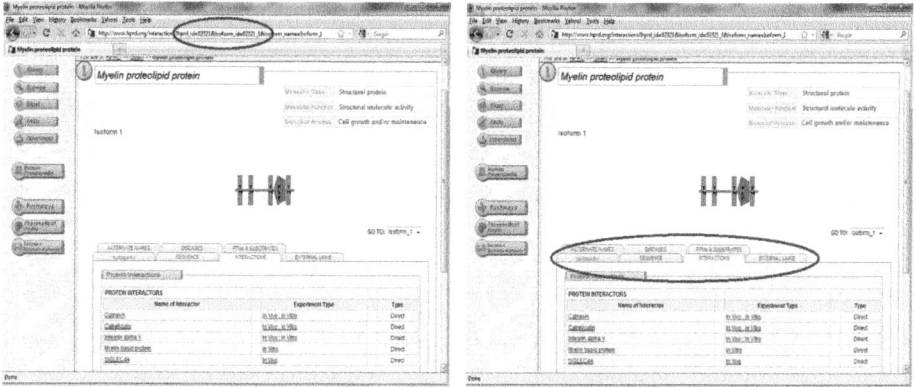

(a) Automatic page redirection.   (b) Incorrect table extraction.

Fig. 9. Multi-page tables and tables with less significant patterns are hard to recognize

The latter issue, however, is not related to Java Scripts, but to how FastWrap dominant pattern recognition algorithm works. Currently FastWrap identifies the largest pattern in the page as the dominant pattern, and thus the target table. It does not take into account that a smaller table may be the object of interest with respect to a given query. To overcome this false recognition, some form of query guided extraction would be more effective.

These examples illustrate the fact that technology migration is also a factor that needs to be addressed. Especially in highly autonomous and distributed systems where one component can change independently making the other that depends on it brittle. Although technologies such as IWebBrowser2 can be used to address some of the problems today's wrapper technologies face, research is required to find a general and robust solution such that systems built using older technologies do not become totally dysfunctional requiring a complete overhaul.

6.2   Optimized Transmission and SQL Injection

We have mentioned earlier that in LifeDB we traded currency for speed and avoided maintaining large data sets as warehouses. The consequence has been that at times, we are forced to ship large amounts of data between sites causing large network latency. Often it is the case, at both client and server ends, that we only need a fraction of the data we trade. Although this issue has not been addressed in the literature adequately, we believe it would be remiss of us to ignore in order to improve online data processing efficiency and to make service composition a viable query processing strategy. In this approach, the goal would be to reduce the data transmission cost by sending only the required portion of the results to the client (mostly). Since services are developed as general purpose solutions, a particular application may not need all the columns and rows that a service will generate. The question is, is it possible to accept selection conditions and projection fields from the client and prepare the desired table before transmission? It turns out that such protocols, though not impossible, pose a serious security issue in form of so called SQL injection [74]. In their work in [74], Roichman and Gudes leveraged our own research on parameterized views [55] to show that SQL views can be used to tailor user needs in a way similar to logic programming rules that can be fired without security breach. We believe the issue is still ill defined and poorly researched, and it may open up a whole new direction of client specific server side query optimization research.

7   Conclusion

Our goal in this paper has been to highlight the features of a new middleware technology for full autonomous large scale data integration in P2P systems, and in the context of this system raise new research issues and challenges that lie ahead toward advancing this technology. The LifeDB system and its query language BioFlow discussed in this paper leverages the developments in the area of schema matching, wrapper generation, key discovery, and automated form filling toward a comprehensive data integration platform. These component technologies are novel research areas in their own right, and it was not our intention to discuss them in this paper. We have discussed the novelty and advantages of LifeDB and BioFlow using only examples on intuitive grounds and refereed the readers to relevant papers for technical details. This is because our major focus has been on presenting the challenges in using these technologies synergistically toward building an autonomous data integration system. These challenges are particularly pronounced when the system is required to have a declarative query interface for users.

The challenges we have highlighted can broadly be classified into two groups – language related and system related. In the language area the main obstacle is how different data types such as DNA sequences, phylogenetic trees, protein interaction, pathways, protein structure, gene expression, annotation, etc. can be managed and manipulated using a single declarative language. The co-mingling of XML and relational data types, and uncertainty of data pose other serious application challenges that we believe need to be addressed. In the system area, the challenges are in developing effective technology blind systems that can withstand technology migration of remote systems that a data integration system might use but have no control over. Apart from technology migration, we have highlighted several technical limitations of current wrapper systems that also need to be removed to improve the query processors in systems such as LifeDB. We believe, addressing the challenges we raised will not only benefit LifeDB and BioFlow, they will also advance the state of the art of the technologies involved and the autonomous data integration in general.

Acknowledgements

This research was supported in part by National Science Foundation grant IIS 0612203 and National Institute of Health grant NIDA 1R03DA026021-01. The authors would also like to acknowledge the insightful discussions with Leonard Lipovich and Fabien Dachet at the Center for Molecular Medicine and Genetics, and Aminul Islam at the Department of Computer Science at Wayne State University during the development of LifeDB that shaped some of its features.

References

1. Affy package,
   http://www.bioconductor.org/packages/2.0/bioc/html/affy.html
2. Affymetrix, http://www.affymetrix.com
3. Agilent Technologies, http://www.agilent.com
4. BAMM Data Set, MetaQuerier Home Page,
   http://metaquerier.cs.uiuc.edu/repository
5. eXist System Home Page, http://exist.sourceforge.net/index.html
6. Illumina, Inc, http://www.illumina.com/
7. LifeDB Data Management System Home Page,
   http://integra.cs.wayne.edu:8080/lifedb
8. MGED Society, http://www.mged.org/Workgroups/MAGE/mage.html
9. OCA Portal, http://www.ebi.ac.uk/msd-srv/oca/oca-docs/oca-home.html
10. The Open Protein Structure Annotation Network, http://www.topsan.org
11. XSB, http://xsb.sourceforge.net/
12. Ahmed, E., Jamil, H.: Post processing wrapper generated tables for labeling anonymous datasets. In: ACM International Workshop on Web Information and Data Management, Hong Kong, China (November 2009)
13. Ala, U., Piro, R.M., Grassi, E., Damasco, C., Silengo, L., Oti, M., Provero, P., Cunto, F.D.: Prediction of human disease genes by human-mouse conserved coexpression analysis. PLoS Comput Biology 4(3), 1–17 (2008)
14. Altintas, I., Berkley, C., Jaeger, E., Jones, M., Ludascher, B., Mock, S.: Kepler: An extensible system for design and execution of scientific workflows. In: SSDBM, p. 423 (2004)
15. Amin, M.S., Jamil, H.: Top-k similar graph enumeration using neighborhood biased β-signatures in biological networks. Technical report, Department of Computer Science, Wayne State University, Detroit, MI, Under review ACM TCBB (July 2010)
16. Amin, M.S., Bhattacharjee, A., Jamil, H.: Wikipedia driven autonomous label assignment in wrapper induced tables with missing column names. In: ACM International Symposium on Applied Computing, Sierre, Switzerland, pp. 1656–1660 (March 2010)
17. Amin, M.S., Bhattacharjee, A., Russell, J., Finley, L., Jamil, H.: A stochastic approach to candidate disease gene subnetwork extraction. In: ACM International Symposium on Applied Computing, Sierre, Switzerland, pp. 1534–1538 (March 2010)
18. Amin, M.S., Jamil, H.: Ontology guided autonomous label assignment for wrapper induced tables with missing column names. In: IEEE International Conference on Information Reuse and Integration, Las Vegas, Nevada (August 2009)
19. Amin, M.S., Jamil, H.: An efficient web-based wrapper and annotator for tabular data. International Journal of Software Engineering and Knowledge Engineering 20(2), 215–231 (2010); IEEE IRI 2009 Special Issue

20. Aumueller, D., Do, H.H., Massmann, S., Rahm, E.: Schema and ontology matching with coma++. In: SIGMOD Conference, pp. 906–908 (2005)
21. Bauckmann, J.: Automatically Integrating Life Science Data Sources. In: VLDB PhD Workshop (2007)
22. Bentley, D.R.: Whole-genome re-sequencing. Current Opinion in Genetics & Development 16(6), 545–552 (2006)
23. Bhattacharjee, A., Islam, A., Amin, M.S., Hossain, S., Hosain, S., Jamil, H., Lipovich, L.: On-the-fly integration and ad hoc querying of life sciences databases using LifeDB. In: 20th International Conference on Database and Expert Systems Applications, Linz, Austria, pp. 561–575 (August 2009)
24. Bhattacharjee, A., Jamil, H.: A schema matching system for autonomous data integration. International Journal of Information and Decision Sciences (2010)
25. Bhattacharjee, A., Jamil, H.: WSM: A novel algorithm for subgraph matching in large weighted graphs. Jouornal of Intelligent Information Systems (to appear, in press, 2011)
26. Bonifati, A., Chang, E.Q., Ho, T., Lakshmanan, L.V.S., Pottinger, R., Chung, Y.: Schema mapping and query translation in heterogeneous p2p xml databases. VLDB J. 19(2), 231–256 (2010)
27. Boulakia, S.C., Biton, O., Davidson, S.B., Froidevaux, C.: Bioguidesrs: querying multiple sources with a user-centric perspective. Bioinformatics 23(10), 1301–1303 (2007)
28. Brazma, A., Hingamp, P., Quackenbush, J., Sherlock, G., Spellman, P., Stoeckert, C., Aach, J., Ansorge, W., Ball, C.A., Causton, H.C., Gaasterland, T., Glenisson, P., Holstege, F.C., Kim, I.F., Markowitz, V., Matese, J.C., Parkinson, H., Robinson, A., Sarkans, U., Schulze-Kremer, S., Stewart, J., Taylor, R., Vilo, J., Vingron, M.: Minimum information about a microarray experiment (MIAME)-toward standards for microarray data. Nature Genetics (December 2001)
29. Chang, C.-H., Lui, S.-C.: Iepad: information extraction based on pattern discovery. In: WWW, pp. 681–688 (2001)
30. Chang, Y.-H., Lee, C.-Z.: Representing multiple mappings between XML and relational schemas for bi-directional query translation. In: Li, Q., Feng, L., Pei, J., Wang, S.X., Zhou, X., Zhu, Q.-M. (eds.) APWeb/WAIM 2009. LNCS, vol. 5446, pp. 100–112. Springer, Heidelberg (2009)
31. Chen, D., Burleigh, G.J., Bansal, M.S., Fernandez-Baca, D.: PhyloFinder: an intelligent search engine for phylogenetic tree databases. BMC Evolutionary Biology 8, 90 (2008)
32. Chen, L., Jamil, H.M.: On using remote user defined functions as wrappers for biological database interoperability. International Journal on Cooperative Information Systems 12(2), 161–195 (2003)
33. Davidson, S.B., Overton, G.C., Tannen, V., Wong, L.: Biokleisli: A digital library for biomedical researchers. Int. J. on Digital Libraries 1(1), 36–53 (1997)
34. Deelman, E., et al.: Pegasus: A framework for mapping complex scientific workflows onto distributed systems. Scientific Programming 13(3), 219–237 (2005)
35. Elhai, J., Taton, A., Massar, J.P., Myers, J.K., Travers, M., Casey, J., Slupesky, M., Shrager, J.: BioBIKE: A web-based, programmable, integrated biological knowledge base. Nucleic Acids Research 37(Web-Server-Issue), 28–32 (2009)
36. Fan, W., Yu, J.X., Li, J., Ding, B., Qin, L.: Query translation from xpath to sql in the presence of recursive dtds. VLDB J. 18(4), 857–883 (2009)

37. Firat, A., Madnick, S.E., Yahaya, N.A., Kuan, C.W., Bressan, S.: Information aggregation using the caméléon# web wrapper. In: Bauknecht, K., Pröll, B., Werthner, H. (eds.) EC-Web 2005. LNCS, vol. 3590, pp. 76–86. Springer, Heidelberg (2005)

38. Funderburk, J.E., Malaika, S., Reinwald, B.: XML programming with SQL/XML and XQuery. IBM Systems Journal 41(4), 642–665 (2002)

39. Gal, A., Modica, G., Jamil, H., Eyal, A.: Automatic ontology matching using application semantics. AI Magazine 26(1), 21–31 (2005)

40. Gamalielsson, J., Olsson, B.: Gosap: Gene ontology-based semantic alignment of biological pathways. IJBRA 4(3), 274–294 (2008)

41. Gusfield, D., Stoye, J.: Relationships between p63 binding, dna sequence, transcription activity, and biological function in human cells. Mol. Cell 24(4), 593–602 (2006)

42. Hancock, D., Wilson, M., Velarde, G., Morrison, N., Hayes, A., Hulme, H., Wood, A.J., Nashar, K., Kell, D., Brass, A.: maxdload2 and maxdbrowse: standards-compliant tools for microarray experimental annotation, data management and dissemination. BMC Bioinformatics 6(1), 264 (2005)

43. He, B., Zhang, Z., Chang, K.C.-C.: Metaquerier: querying structured web sources on-the-fly. In: SIGMOD Conference, pp. 927–929 (2005)

44. He, H., Singh, A.K.: Graphs-at-a-time: query language and access methods for graph databases. In: SIGMOD Conference, pp. 405–418 (2008)

45. Hoon, S., Ratnapu, K.K., Chia, J.-M., Kumarasamy, B., Juguang, X., Clamp, M., Stabenau, A., Potter, S., Clarke, L., Stupka, E.: Biopipe: A flexible framework for protocol-based bioinformatics analysis. Genome Research 13(8), 1904–1915 (2003)

46. Hosain, S., Jamil, H.: An algebraic foundation for semantic data integration on the hidden web. In: Third IEEE International Conference on Semantic Computing, Berkeley, CA (September 2009)

47. Hossain, S., Islam, M., Jesmin, Jamil, H.: PhyQL: A web-based phylogenetic visual query engine. In: IEEE Conference on Bioinformatics and BioMedicine, Philadelphia, PA (November 2008)

48. Hossain, S., Jamil, H.: A visual interface for on-the-fly biological database integration and workflow design using VizBuilder. In: 6th International Workshop on Data Integration in the Life Sciences (July 2009)

49. Hull, D., Wolstencroft, K., Stevens, R., Goble, C., Pocock, M.R., Li, P., Oinn, T.: Taverna: a tool for building and running workflows of services. Nucleic Acids Res., 34 (2006), Web Server issue

50. Jahnkuhn, H., Bruder, I., Balouch, A., Nelius, M., Heuer, A.: Query transformation of SQL into xQuery within federated environments. In: Grust, T., Höpfner, H., Illarramendi, A., Jablonski, S., Fischer, F., Müller, S., Patranjan, P.-L., Sattler, K.-U., Spiliopoulou, M., Wijsen, J. (eds.) EDBT 2006. LNCS, vol. 4254, pp. 577–588. Springer, Heidelberg (2006)

51. Jamil, H.: A novel knowledge representation framework for computing sub-graph isomorphic queries in interaction network databases. In: International Conference on Tools with Artificial Intelligence, Newark, NJ, pp. 131–138 (November 2009)

52. Jamil, H.: Computing subgraph isomorphic queries using structural unification and minimum graph structures. In: ACM International Symposium on Applied Computing, Taichung, Taiwan (March 2011)

53. Jamil, H., Islam, A.: Managing and querying gene expression data using Curray. BMC Proceedings 5(suppl. 2), S10 (2011)

54. Jamil, H., Islam, A., Hossain, S.: A declarative language and toolkit for scientific workflow implementation and execution. International Journal of Business Process Integration and Management 5(1), 3–17 (2010); IEEE SCC/SWF 2009 Special Issue on Scientific Workflows

55. Jamil, H.M.: A case for parameterized views and relational unification. In: ACM International Symposium on Applied Computing, pp. 275–279 (2001)

56. Jin, L., Li, C., Mehrotra, S.: Efficient record linkage in large data sets. In: DASFAA, p. 137 (2003)

57. Krishnaprasad, M., Liu, Z.H., Manikutty, A., Warner, J.W., Arora, V.: Towards an industrial strength SQL/XML infrastructure. In: ICDE (2005)

58. Laender, A.H.F., Ribeiro-Neto, B., da Silva, A.S.: Debye - date extraction by example. Data Knowl. Eng. 40(2), 121–154 (2002)

59. Lakshmanan, L.V.S., Sadri, F.: On a theory of probabilistic deductive databases. TPLP 1(1), 5–42 (2001)

60. Lakshmanan, L.V.S., Sadri, F.: On the information content of an XML database. Manuscript. University of North Carolina at Greensboro (2009), http://www.uncg.edu/~sadrif/ict-full.pdf

61. Lakshmanan, L.V.S., Sadri, F., Subramanian, I.N.: Schemasql - a language for interoperability in relational multi-database systems. In: VLDB Proceedings, pp. 239–250 (1996)

62. Lakshmanan, L.V.S., Shiri, N.: A parametric approach to deductive databases with uncertainty. IEEE Trans. Knowl. Data Eng. 13(4), 554–570 (2001)

63. Lee, M.-L., Yang, L.H., Hsu, W., Yang, X.: Xclust: clustering xml schemas for effective integration. In: CIKM, pp. 292–299 (2002)

64. Lengu, R., Missier, P., Fernandes, A.A.A., Guerrini, G., Mesiti, M.: Time-completeness trade-offs in record linkage using adaptive query processing. In: EDBT, pp. 851–861 (2009)

65. Li, G., Kou, G.: Aggregation of information resources on the invisible web. In: WKDD, pp. 773–776 (2009)

66. Li, H., Coghlan, A., Ruan, J., Coin, L.J.J., Hériché, J.-K.K., Osmotherly, L., Li, R., Liu, T., Zhang, Z., Bolund, L., Wong, G.K.-S.K., Zheng, W., Dehal, P., Wang, J., Durbin, R.: TreeFam: a curated database of phylogenetic trees of animal gene families. Nucleic Acids Research 34(Database issue) (2006)

67. Li, P., Castrillo, J.I., Velarde, G., Wassink, I., Soiland-Reyes, S., Owen, S., Withers, D., Oinn, T., Pocock, M.R., Goble, C.A., Oliver, S.G., Kell, D.B.: Performing statistical analyses on quantitative data in taverna workflows: an example using r and maxdBrowse to identify differentially-expressed genes from microarray data. BMC Bioinformatics 9(1), 334 (2008)

68. Madhavan, J., Bernstein, P.A., Rahm, E.: Generic schema matching with cupid. In: VLDB, pp. 49–58 (2001)

69. Madria, S.K., Passi, K., Bhowmick, S.S.: An xml schema integration and query mechanism system. Data Knowl. Eng. 65(2), 266–303 (2008)

70. Majithia, S., Shields, M., Taylor, I., Wang, I.: Triana: A graphical web service composition and execution toolkit. In: IEEE ICWS, p. 514 (2004)

71. Michelson, M., Knoblock, C.A.: Learning blocking schemes for record linkage. In: AAAI (2006)

72. Quackenbush, J.: Computational approaches to analysis of dna microarray data. Yearbook of Medical Informatics 1, 91–103 (2006)

73. R Development Core Team. R: A Language and Environment for Statistical Computing. R Foundation for Statistical Computing, Vienna, Austria (2009) ISBN 3-900051-07-0

74. Roichman, A., Gudes, E.: Fine-grained access control to web databases. In: SAC-MAT, pp. 31–40 (2007)
75. Sadri, F.: Information source tracking method: Efficiency issues. IEEE Trans. Knowl. Data Eng. 7(6), 947–954 (1995)
76. Sismanis, Y., Brown, P., Haas, P.J., Reinwald, B.: GORDIAN: efficient and scalable discovery of composite keys. In: VLDB 2006, pp. 691–702 (2006)
77. Sultana, K.Z., Bhattacharjee, A., Jamil, H.: IsoKEGG: A logic based system for querying biological pathways in KEGG. In: IEEE International Conference on Bioinformatics and Biomedicine (December 2010)
78. Ullmann, J.R.: An algorithm for subgraph isomorphism. Journal of ACM 23(1), 31–42 (1976)
79. Valiente, G.: Algorithms on Trees and Graphs. Springer, Berlin (2002)
80. Wang, K., Tarczy-Hornoch, P., Shaker, R., Mork, P., Brinkley, J.: Biomediator data integration: Beyond genomics to neuroscience data. In: AMIA Annu. Symp. Proc., pp. 779–783 (2005)
81. Wernicke, S., Rasche, F.: Simple and fast alignment of metabolic pathways by exploiting local diversity. Bioinformatics 23(15), 1978–1985 (2007)
82. Yu, J., Finley, R.: Combining multiple positive training sets to generate confidence scores for protein–protein interactions. Bioinformatics 25(1), 105–111 (2009)
83. Yu, J., Pacifico, S., Liu, G., Finley, R.: Droid: the drosophila interactions database, a comprehensive resource for annotated gene and protein interactions. BMC Genomics 9(1), 461 (2008)
84. Zhang, J.D., Wiemann, S.: KEGGgraph: A graph approach to KEGG PATHWAY in R and Bioconductor. Bioinformatics (March 2009)
85. Zhang, Y., Boncz, P.: XRPC: interoperable and efficient distributed XQuery. In: VLDB, pp. 99–110 (2007)
86. Rose, P.W., Beran, B., Bi, C., Bluhm, W., Dimitropoulos, D., Goodsell, D.S., Prlic, A., Quesada, M., Quinn, G.B., Westbrook, J.D., Young, J., Yukich, B.T., Zardecki, C., Berman, H.M., Bourne, P.E.: The RCSB Protein Data Bank: redesigned web site and web services. Nucleic Acids Research 39(Database-Issue), 392–401 (2011)

Author Index